PROJECT AND COST ENGINEERS' HANDBOOK
Fourth Edition

COST ENGINEERING

A Series of Reference Books and Textbooks

Editor

KENNETH K. HUMPHREYS, Ph.D.
Consulting Engineer

Granite Falls, North Carolina

PROJECT AND COST ENGINEERS' HANDBOOK
Fourth Edition

edited by
KENNETH K. HUMPHREYS
Consulting Engineer
Granite Falls, North Carolina, U.S.A.

 Marcel Dekker   New York

Library of Congress Cataloging-in-Publication Data
A catalog record for this book is available from the Library of Congress.

ISBN: 0-8247-5746-7

This book is printed on acid-free paper.

Headquarters
Marcel Dekker, 270 Madison Avenue, New York, NY 10016, U.S.A.
tel: 212-696-9000; fax: 212-685-4540

Distribution and Customer Service
Marcel Dekker, Cimarron Road, Monticello, New York 12701, U.S.A.
tel: 800-228-1160; fax: 845-796-1772

World Wide Web
http://www.dekker.com

Current printing (last digit):
10 9 8 7 6 5 4 3 2 1

PRINTED IN THE UNITED STATES OF AMERICA

Dr. James M. Neil, PE, 1927–2003

Jim Neil served the cost engineering profession with distinction as an author, educator, and professional engineer. A strong supporter of education, Jim Neil worked tirelessly with the AACE International Education Board in developing reference materials for the profession and in fostering the growth of and support for scholarship programs. Upon Jim Neil's unexpected passing in 2003, Barry McMillan, Executive Director of AACE International said, "Jim quietly made an enormous impact on our Association, on the profession, on the business community, and on those who knew him in other walks of life."

This book is dedicated in memory of Dr. James M. Neil, PE. His contributions to the cost engineering profession and to education will long be remembered.

Preface

Twelve years have passed since the publication of the Third Edition of the *Project and Cost Engineers' Handbook.* In some ways nothing about the topic has changed. In other ways, everything has changed. Nothing has changed in the fundamental principles of cost engineering. They remain as they were back then. However, with the explosive growth of e-mail and the Internet, the gathering and sharing of data and information has indeed changed dramatically. E-mail was a novelty when the Third Edition was published, and the fax was the predominant method of sharing documents and information over great distances. Now e-mail, scanners, and cell phones dominate communications and the facsimile machine is destined to become a museum piece.

When the Third Edition was published, "INTERNET" was understood in the cost engineering and project management professions to mean the International Project Management Association. Now "INTERNET" is the worldwide web and the International Project Management Association has, of necessity, abandoned the Internet name and now is simply IPMA.

In this same period IPMA and the International Cost Engineering Council (ICEC), both of which are confederations of national cost and project management associations, have expanded dramatically to include well over 60 national cost and project associations around the world, more than double the previous number of national associations within the IPMA and ICEC organizations.

Projects and products have become more and more multinational in nature, and cost engineers and project managers are increasingly concerned about international cost considerations.

In this same period projects seemingly have become more risky and cost and project engineers are increasingly concerned about risk analysis. Risks have even been exacerbated by the proliferation of cost and schedule software which is all too often misused by persons lacking the experience which is necessary to discern when the output of the software, be it an estimate or a schedule, is realistic. Work which was previously done by an experienced cost or project engineer all too often is now handled by an inexperienced person who relies on software without really understanding its limitations or if the inherent assumptions made when the software was developed are reasonable for the project at hand.

Further, business scandals and corporate mismanagement seem to have proliferated with increased frequency, resulting in project failures, bankruptcies, corporate collapse, and in some cases, criminal indictments. This has brought increased concern about business and engineering ethics.

For these various reasons, except for necessary factual and reference updating and some expansion of Chapter 3, "Cost Estimating," all of the chapters of the Third Edition have been retained in this Edition. However, the Fourth Edition expands on the Third Edition with the addition of discussions of international project considerations, project risk analysis and contingency, ethical considerations and, of course, the Internet.

As was the case with the previous three editions of the *Project and Cost Engineers' Handbook*, this work would not have been possible without the many contributions of members of AACE International, the Association for the Advancement of Cost Engineering. My thanks go to them and to AACE International for once again agreeing to sponsor this book. Special thanks go to the Houston Gulf Coast Section of AACE International and the late Aaron Cohen, PE CCE and the late Thomas C. Ponder, PE CCE who in 1979 were responsible for originating the first edition and to Dr.Lloyd M. English, PE who collaborated with me on the third edition.

Kenneth K. Humphreys, PhD, PE CCE

Acknowledgments

The editor wishes to acknowledge the many individuals who have contributed to all four editions of the *Project and Cost Engineers' Handbook*.

Special thanks are due to AACE International (the Association for the Advancement of Cost Engineering) for once again sponsoring this book as they did with the prior three edition. Thanks are also due to the International Cost Engineering Council and its member associations for providing much of the information which forms the basis for the new material in this edition.

Personal thanks go to Charles P. (Chick) Woodward for his insights on international project considerations and to Michael Curran, President of Decision Sciences Corporation, St. Louis, Missouri, from whom I gained a deep appreciation for concepts of risk analysis and its value in project evaluation.

Contents

Contributors

Dorothy J. Burton[1] *Management Systems Associates, Ltd., St. Louis, Missouri, U.S.A.*

R. L. Dodds[2] *Project Execution Planners, Houston, Texas, U.S.A.*

Lloyd M. English, PE[1] *AACE International, Morgantown, West Virginia, U.S.A.*

Jerry L. Hamlin, PE CCE[2] *Grand Bassa Tankers, Inc., Tulsa, Oklahoma, U.S.A.*

James W. Higgenbotham[2] *St. Regis Paper Company, Houston, Texas, U.S.A.*

Michael E. Horwitz, PE CCE[1] *Simons-Eastern Consultants, Inc., Atlanta, Georgia, U.S.A.*

Kenneth K. Humphreys, PhD PE CCE[3] *Consulting Engineer, Granite Falls, North Carolina, U.S.A.*

Barry G. McMillan[1] *AACE International, Morgantown, West Virginia, U.S.A.*

John P. Nolan, Sr.[1] *Sargent & Lundy Engineers, Chicago, Illinois, U.S.A.*

John J. O'Driscoll[2] *Shell Oil Company, Houston, Texas, U.S.A.*

F. Fred Rahbar[1] *M. M. Al-Rumaih Compound, Jedddah, Saudi Arabia*

E. E. Stackhouse[2] *ICARUS Corporation, Houston, Texas, U.S.A.*

---

[1]Contributors to the third edition and affiliations at that time. The work of these contributors has been retained in the fourth edition.

[2]Contributors to second edition and affiliations at that time. Portions of the work of these contributors appear in the fourth edition.

[3]Contributor to the second, third, and fourth editions. Portions of his prior work appear in the fourth edition.

Wilfred Stelly, CCE[2] *C-E Lummus Co., Houston, Texas, U.S.A.*

Kul B. Uppal, PE[1] *Amoco Chemical Corporation, Houston, Texas, U.S.A.*

A. J. Wallace[2] *C-E Lummus Co., Houston, Texas, U.S.A.*

Richard E. Westney, PE[1] *Spectrum Consultants International, Inc., Houston, Texas, U.S.A.*

Charles P. Woodward, PE CCE *Burns & Roe Services Inc., Virginia Beach, Virginia, U.S.A.*

1

Cost Engineering Basics

This chapter touches on certain basic subjects that are necessary tools of the cost engineer. Many of the topics that are described only briefly are treated later in somewhat greater detail; hence brief treatment of a topic in this chapter does not necessarily indicate that the particular topic is not an essential subject for further study. Chapter 1 is a rather thorough overview of cost engineering as a whole and is meant to be a foundation for the chapters that follow.

For those interested in preparing for any of the AACE International (Association for the Advancement of Cost Engineering) certification examinations (Certified Cost Engineer, Certified Cost Consultant, Interim Cost Consultant, and Planning and Scheduling Professional), Appendix A provides a guide to the examinations and some sample examination questions. This chapter reviews many of the topics which are ordinarily covered in the certification examinations. Terminology used throughout this chapter and subsequent chapters is in accord with that recommended by the AACE International in the 2003 revision of AACE Standard No. 10S-90. This standard is a revision and expansion upon American National Standards Institute Standard No. Z94.2-1989, "Cost Engineering Terminology." An abridged version of this standard is included in this handbook as Appendix B.

1.1 MATHEMATICS OF COST COMPARISONS

The mathematics of cost comparisons involve:

- Time value of money
- Relationships between kinds of cost
- Methods for handling unequal durations

Time has value. Therefore equal quantities of money received (or spent) at different times do not have equal value; to be added or compared accurately they

1

must be reduced to a common time. If a quantity of money, P, earns interest i (expressed in decimal form) for a specified time period, its value at the end of that time period becomes:

$$S_1 = P(1 + i) \tag{1.1}$$

where S_1 equals the future value after one time period. Extending this same equation for n time periods we obtain:

$$S_1 = P(1 + i)^{\pm n} \tag{1.2}$$

The most common time period used is the year, although any time period can be used for compounding interest. The exponent is positive if time is moving with the calendar and negative if it is moving against the calendar. Thus Eq. (1.2) can be used to calculate the present value of either cash received or debt incurred at some future time.

There are four principal timing conventions used for cost evaluations:

P: present value, a single amount
S: future value, a single amount
R: a uniform end-of-year amount
R_b: a uniform beginning-of-year amount

It is convenient to express the relationships between these values with a factor, designated F, with subscripts indicating the conversion being made. Thus:

$F_{RP,i,n}$ is the factor that converts \$1 of R to a present worth P at an interest rate (expressed as a decimal) of i per period for n periods.

$F_{SR,i,n}$ is the factor that converts \$1 of future amount S received n periods from now to a uniform end-of-period amount R at a decimal interest rate of i.

A summary of the relationships along with the algebraic equivalents is given in Table 1.1.

Not only must costs be compared at equivalent points of time, unequal costs of unequal duration must be compared on an equivalent duration basis. The two methods most commonly used for obtaining a common denominator for unequal periods of duration are:

1.   *Unacost*: an equivalent end-of-year cost
2.   *Capitalized Cost*: an equivalent present cost for infinite service

Table 1.1 Summary of S, P, R, and R_b Relationships

Item	Conversion	Algebraic relationship	Relationship by factor	Name of factor
1	P to S	$S = P[(1+i)^n]$	$S = PF_{PS,i,n}$	Compound-interest factor
2	S to P	$P = S[(1+i)^{-n}]$	$P = SF_{SP,i,n}$	Present-value factor
3	R to P	$P = R\dfrac{(1+i)^n - 1}{i(1+i)^n}$	$P = RF_{RP,i,n}$	Unacost present-value factor
4	P to R	$R = P\dfrac{i(1+i)^n}{(1+i)^n - 1}$	$R = PF_{PR,i,n}$	Capital-recovery factor
5	R to S	$S = R\dfrac{(1+i)^n - 1}{i}$	$S = RF_{RS,i,n}$ $S = RF_{RP,i,n}F_{PS,i,n}$	Equal-payment-series future-value factor
6	S to R	$R = S\dfrac{i}{(1+i)^n - 1}$	$R = SF_{SR,i,n}$ $R = SF_{SP,i,n}F_{PR,i,n}$	Sinking-fund factor
7	R_b to P	$P = R_b(1+i)\dfrac{(1+i)^n - 1}{i(1+i)^n}$	$P = R_b(1+i)F_{RP,i,n}$	
8	P to R_b	$R_b = \dfrac{P}{1+i}\dfrac{i(1+i)^n}{(1+i)^n - 1}$	$R_b = \dfrac{P}{1+i}F_{PR,i,n}$	
9	R_b to S	$S = R_b(1+i)\dfrac{(1+i)^n - 1}{i}$	$S = R_b(1+i)F_{RS,i,n}$ $S = R_b(1+i)F_{RP,i,n}F_{PS,i,n}$	
10	S to R_b	$R_b = \dfrac{S}{1+i}\dfrac{i}{(1+i)^n - 1}$	$R_b = S\dfrac{1}{1+i}F_{SR,i,n}$ $R_b = S\dfrac{1}{1+i}F_{SP,i,n}F_{PR,i,n}$	

Source: K. K. Humphreys, *Jelen's Cost and Optimization Engineering*, 3rd ed., 1991, McGraw-Hill, Inc., New York. p. 29. Reprinted by permission.

The prefix "una-" designates a "uniform annual" relationship. Thus unacost is a uniform annualized cost. Unacost values are denoted by R, and capitalized costs are denoted by K. They are related by the time value of money:

$$R = iK \tag{1.3}$$

Both unacost and capitalized cost are merely mathematical conveniences for obtaining common denominators to compare service lives. There is no assumption that a 1-year life or a perpetual life is required. A summary of unacost and capitalized cost relationships is given in Table 1.2.

1.2   DEPRECIATION

In practice, cost comparisons are made on an after-tax basis. Depreciation must be considered in determining the tax base on which taxes are paid. Depreciation arises when an expenditure lasts for more than 1 year. A depreciation schedule is a means of allocating the expenditure as an expense over the years; the accountant's application of depreciation for each year is called a depreciation expense account. Depreciation is an expense for tax purposes. There are numerous methods of establishing depreciation schedules in use throughout the world. Five commonly used techniques are described below. Although these techniques are widely used in many countries, because of several tax law revisions since 1981 they now have limited applicability in the United States.

1. *Straight-line depreciation* (*SL*). The depreciation is constant for all the years and is for each year

$$\frac{C_d}{n} \tag{1.4}$$

where C_d is the depreciable first cost and n is the depreciable life in years.
2. *Declining-balance depreciation* (*DB*). A factor F_{DB} is determined from the salvage value and the life and is used as follows to obtain a decreasing depreciation for the individual years:

$$F_{DB} = 1 - n\sqrt{\frac{C_{sal}}{C_i}} \tag{1.5}$$

where C_i is the initial cost. The depreciation for the years is obtained from the following sequence:

$$D_1 = C_i F_{DB}$$
$$D_2 = C_i(1 - F_{DB})F_{DB}$$
$$D_3 = C_i(1 - F_{DB})^2 F_{DB}$$

$$\vdots$$

Table 1.2 Summary of Unacost and Capitalized Cost Relationships—No Tax

Item no.	Item	Present value as a cost for n years' duration	Unacost		Capitalized cost	
			Algebraic expression	Factor expression	Algebraic expression	Factor expression
1	C_i Initial cost for n-year life	C_i	$C_i\dfrac{i(1+i)^n}{(1+i)^n-1}$	$C_iF_{PR,i,n}$	Multiply corresponding algebraic expression for unacost by $\dfrac{1}{i}$	$C_iF_{PK,i,n}$
2	R Uniform end-of-year annual cost	$R\dfrac{(1+i)^n-1}{i(1+i)^n}$	R	$R1$ (factor is unity)		$R\dfrac{1}{i}$
3	R_b Uniform beginning-of-year annual cost	$R_b(1+i)\dfrac{(1+i)^n-1}{i(1+i)^n}$	$R_b(1+i)$	$R_b(1+i)$		$R_bF_{PK,i,1}$
4	C_{ex} Irregular cost at end of xth year for article lasting n years	$C_{ex}\dfrac{1}{(1+i)^x}$	$C_{ex}\dfrac{1}{(1+i)^x}\times\dfrac{i(1+i)^n}{(1+i)^n-1}$	$C_{ex}\dfrac{1}{(1+i)^x}F_{PR,i,n}$		$C_{ex}\dfrac{1}{(1+i)^x}F_{PK,i,n}$

(continued)

Table 1.2 *(Continued)*

Item no.	Item	Present value as a cost for n years' duration	Unacost — Algebraic expression	Unacost — Factor expression	Capitalized cost — Algebraic expression	Capitalized cost — Factor expression
5	C_{bx} Irregular cost at beginning of xth year for article lasting n years	$C_{bx}\dfrac{1}{(1+i)^{x-1}}$	$C_{bx}\dfrac{1}{(1+i)^{x-1}}\times\dfrac{i(1+i)^n}{(1+i)^n-1}$	$C_{bx}\dfrac{1}{(1+i)^{x-1}}F_{PR,i,n}$	Multiply corresponding algebraic expression for unacost by $\dfrac{1}{i}$	$C_{bx}\dfrac{1}{(1+i)^{x-1}}F_{PK,i,n}$
6	C_{nd} Nondepreciable first cost, e.g., land or an article that lasts forever	$C_{nd}\dfrac{(1+i)^n-1}{(1+i)^n}$	$C_{nd}i$	$C_{nd}i$		$C_{nd}1$ (factor is unity)
7	C_{sal} Salvage value at end of nth year, a receipt	$-C_{sal}\dfrac{1}{(1+i)^n}$	$-C_{sal}\dfrac{i}{(1+i)^n-1}$	$-C_{sal}\dfrac{1}{(1+i)^n}F_{PR,i,n}$		$-C_{sal}(F_{PR,i,n}-1)$ or $-C_{sal}\dfrac{1}{(1+i)}F_{PR,i,n}$

Source: K. K. Humphreys, *Jelen's Cost and Optimization Engineering*, 3nd ed., 1991, McGraw-Hill, Inc., New York, pp. 40–41. Reprinted by permission.

3. *Sum-of-the-years-digits depreciation (SD)*. The general formula takes the form:

$$D_m = C_i\left(\frac{n - m + 1}{0.5n(n + 1)}\right) \qquad (1.6)$$

where D_m is the depreciation in year m. This method is best illustrated by the example of a purchase that has a 4-year depreciable life. The sum-of-the-years-digits is $4 + 3 + 2 + 1 = 10$. The depreciation per year of an initial depreciable cost C_d is:

$$D_1 = C_d\frac{4}{10}$$

$$D_2 = C_d\frac{3}{10}$$

$$D_3 = C_d\frac{2}{10} \qquad (1.7)$$

$$D_4 = C_d\frac{1}{10}$$

where $C_d = C_i - C_{\text{sal}}$. Obviously, the yearly depreciable cost decreases with time.

4. *Double-declining-balance depreciation (DDB)*. The method resembles the declining-balance method, but the factor F_{DDB} is obtained from

$$F_{\text{DDB}} = \frac{2}{n} \qquad (1.8)$$

and the decreasing depreciation for the years is

$$D_1 = C_i F_{\text{DDB}}$$
$$D_2 = C_i(1 - F_{\text{DDB}})F_{\text{DDB}}$$

$$D_3 = C_i(1 - F_{\text{DDB}})^2 F_{\text{DDB}}$$

$$\vdots$$

$$D_n = C_i(1 - F_{\text{DDB}})^{n-1} F_{\text{DDB}}$$

Although the double-declining-balance method requires a factor $2/n$, it is permissible to select any smaller factor, such as $1.5/n$. It is also permissible to combine assets in a group, and such grouping is widely used to avoid an unnecessary burden of a profusion of accounts.

Table 1.3 Relationships for Depreciation, n-Year Life[a]

Item no.	Method	Annual depreciation for mth year	Accumulated depreciation for m years	Present value of m years of depreciation
1	Straight-line (SL)	$\dfrac{C_i - C_{sal}}{n}$	$(C_i - C_{sal})\dfrac{m}{n}$	$\dfrac{C_i - C_{sal}}{n}F_{RP,i,n}$
2	Sinking-fund (SF)	$(C_i - C_{sal})F_{SP,i,n}F_{PR,i,n}F_{PS,i,m-1}$	$(C_i - C_{sal})\dfrac{F_{PS,i,m}-1}{F_{PS,i,n}-1}$	$(C_i - C_{sal})\dfrac{m}{1+i}F_{SP,i,n}F_{PR,i,n}$
3	Declining-balance (DB)	$C_i(1-F_{DB})^{m-1}F_{DB}$	$C_i[1 - (1-F_{DB})^m]$	$C_iF_{DB}\dfrac{1-\left(\dfrac{1-F_{DB}}{1+i}\right)^m}{i+F_{DB}}$
4	Sum-of-the-years-digits (SD)	$(C_i - C_{sal})\dfrac{n-m+1}{0.5n(n+1)}$	$(C_i - C_{sal})\dfrac{m}{n(n+1)}(2n+1-m)$	$\dfrac{2(C_i - C_{sal})}{n(n+1)i}$ $\times[n - F_{RP,i,m} - (n-m)F_{SP,i,m}]$
5	Units-of-production (UP)	$(C_i - C_{sal})\dfrac{M_m}{m}$	$(C_i - C_{sal})\dfrac{\sum_1^m M_m}{M}$	$\dfrac{C_i - C_{sal}}{M}\sum_1^m\dfrac{M_m}{(1+i)^m}$

[a]Book value at end of mth year = C_i less accumulated depreciation for m years. Depreciation remaining for $n - m$ future years = book value at end of mth year less C_{sal}. For declining-balance depreciation, terminal value is not related to salvage value, but total write-off should be to salvage value. At the end of useful life, $m = n$.

Source: K. K. Humphreys, Jelen's Cost and Optimization Engineering, 3rd ed., 1991, McGraw-Hill, Inc., New York, p. 62. Reprinted by permission.

5. *Units of Production (UP)*. This form of depreciation is used for a capital acquisition that has a distinct productive capacity (M) over its life. In this method

$$D_m = C_d\left(\frac{M_m}{M}\right) \tag{1.9}$$

where M_m = the production in year m. For example, a drilling rig having no salvage value that was purchased with an assumed potential of drilling 250,000 total feet would be depreciated by 10% of its initial cost in a year in which it drilled 25,000 feet; 20% in a year that it drilled 50,000 feet; etc.

Relationships for these various depreciation schedules are given in Table 1.3.

The different depreciation methods offer different procedures for allocating the depreciation over the individual years; total allowable depreciation for tax purposes is the same for all methods. In no case may the book value fall below the predetermined salvage value, as this would amount to excessive depreciation. Straight-line depreciation is a uniform method. Units of production depreciation is related solely to production and may rise or fall from year to year. The other methods mentioned are accelerated methods, with the rate of depreciation highest at the start. The double-declining-balance method gives the highest present value for the future depreciation followed by the sum-of-the-years-digits method, the declining-balance method, and the straight-line method. A sinking fund depreciation method, not discussed here, has an increasing rate of depreciation with time but has not generally been used by industry.

1.3 ACCELERATED COST RECOVERY SYSTEM: 1981–1986

In 1981 a major revision of tax laws in the United States replaced the preexisting depreciation systems described above with a system known as the Accelerated Cost Recovery System (ACRS). ACRS was mandatory in the United States for all capital assets acquired from 1981 to 1986. Then, in 1987 ACRS was replaced by but another mandatory system, the Modified Accelerated Cost Recovery System (MACRS). These new systems did not apply to capital assets acquired prior to 1981 and thus those older assets continued to be depreciated according to the system in use at the time they were acquired. Since many of the older assets had depreciable lives of 50 years or more, the old depreciation systems will be used by cost professionals in the United States at least until 2030.

Under ACRS, capital assets were not subject to depreciation in the customary sense. It was not necessary to estimate salvage values or useful lives for equipment. Instead the law established various property classes and provided for deductions calculated as specific percentages of the cost of the asset. The classes established were in 6 categories: 3-, 5-, 10-, and 15-year properties and

10- and 15-year real properties. For detailed definitions of each class, see IRS Publication 534, *Depreciation Property Placed in Service Before 1987.* Because most types of industrial equipment and machinery fell into the 5-year category and because the longest ACRS class life was specified as 15 years, ACRS is now generally only of historic interest for cost engineers.

Under the 1981 law, ACRS deductions were phased in on a gradual basis over a period of years. Fifteen-year class real property deductions varied and were based in part on the month in which the property was placed in service.

It should also be noted that the ACRS requirements placed certain limitations on deductions for disposition of property prior to the end of its class life, and also provided for optional use of alternate percentages based on the straight-line method of depreciation. If the optional method was chosen to depreciate an asset or if the asset was real property, it is possible that the asset is still subject to ACRS depreciation rules.

1.4 MODIFIED ACCELERATED COST RECOVERY SYSTEM (MACRS)

In the United States, all assets acquired after December 31, 1986 must be depreciated under the MACRS system, as established by the U.S. Tax Reform Act of 1986, and as modified by subsequent tax law changes. Any property depreciated under MACRS is assigned a recovery period based upon its class life. A class life is the expected productive or usable life of a piece of property as defined by the Internal Revenue Service (IRS); the recovery period is the length of time that the IRS will allow for depreciation. Class life and recovery period generally differ from each other. For example, most industrial process equipment is assigned a recovery period of 7 years although it has a class life ranging from 10 to 16 years. Deductions are calculated as a specific percentage of original cost according to the recovery period. As with ACRS, the salvage value is not considered in the calculations.

MACRS, now recognizes 6 property classifications (*not* class lives), ranging from 3-year to 20-year property, plus classifications for water utility property, residential real property, nonresidential real property, and certain railroad property improvements, all of which have specified depreciation periods in excess of 20 years. Straight-line depreciation applies to these classifications.

The distinctions between MACRS classifications are not perfectly clear and can be somewhat ambiguous. Although general definitions are available in various sources, the only way to properly classify property is to consult the IRS directly as required.

New technology is not covered under existing property classifications. In the event that a new technology is put into place that has no precedent for tax purposes, the IRS must be contacted, and a decision will be made by the IRS to assign a specific recovery period. It is a good practice for the applicant to be prepared at this point with a specific recommendation and with data to support that recommendation. The first ruling on a new technology will set a precedent.

A half-year convention normally applies when using MACRS; that is, it is assumed that all acquisitions in a given year are made in midyear, thus yielding one-half of the full-year depreciation in the first year of an asset's life. Under special circumstances, a mid-quarter convention may apply.

Table 1.4 shows the MACRS allowances for depreciation for the various property classes up to 20-years. Property in the 3-, 5-, 7-, and 10-year classes are depreciated under the double (200%)-declining-balance method. Property in the 15- and 20-year classes are depreciated with the 150% Declining Balance method. In both cases, as soon as the calculated declining balance deduction falls below the straight-line method, the depreciation schedule is switched over to the latter. As stated above, straight-line depreciation applies to classifications with recovery periods greater than 20 years.

The option still exists to depreciate most 3- to 20-year property over its class life with straight-line depreciation, but the half-year convention must still be followed. A midmonth convention applies to the classifications greater than 20-years.

There are other options and provisions that are not discussed here. Any tax regulation is always subject to revision. Therefore always obtain the most recent tax information available from the appropriate government agency, federal, state, or local, before finalizing any financial planning or estimation. (Note: Property taxes and similar taxes not based on income are classified as operating expenses and are thus not considered in this chapter.)

1.5 COST COMPARISONS OF UNEQUAL DURATIONS

As mentioned previously, unacost and capitalized cost are used for comparison of costs of unequal duration. In making the comparison, after-tax values should be considered. All depreciation, of course, has a net present value that is not only dependent upon the time value of money, but also on the prevailing tax rate. Thus $1 of the depreciable first cost of a piece of 5-year property has the following cost diagram:

$$D_1 t \quad D_2 t \quad D_3 t \quad D_4 t \quad D_5 t$$

$1

where D is the fractional depreciation for each year and t is the tax rate. At each year end the tax base is reduced by D, the depreciation expense, and Dt represents the tax benefit. Thus the present value of the benefit is:

$$P = 1 - \sum \frac{D_n t}{(1+r)^n} \tag{1.10}$$

Table 1.4   MACRS Deduction Rates

General depreciation system
Applicable depreciation method: 200 or 150 percent
Declining balance switching to straight line
Applicable recovery periods: 3, 5, 7, 10, 15, 20 years
Applicable convention: Half-year

If the recovery year is:	And the recovery period is:					
	3-year	5-year	7-year	10-year	15-year	20-year
	The depreciation rate is:					
1	33.33	20.00	14.29	10.00	5.00	3.750
2	44.45	32.00	24.49	18.00	9.50	7.219
3	14.81	19.20	17.49	14.40	8.55	6.677
4	7.41	11.52	12.49	11.52	7.70	6.177
5		11.52	8.93	9.22	6.93	5.713
6		5.76	8.92	7.37	6.23	5.285
7			8.93	6.55	5.90	4.888
8			4.46	6.55	5.90	4.522
9				6.56	5.91	4.462
10				6.55	5.90	4.461
11				3.28	5.91	4.462
12					5.90	4.461
13					5.91	4.462
14					5.90	4.461
15					5.91	4.462
16					2.95	4.461
17						4.462
18						4.461
19						4.462
20						4.461
21						2.231

Note: 200 percent declining balance applies to class lives of 10 years or less; 150 percent applies to 15- and 20-year class lives.
Source: US Internal Revenue Service, *Instructions for Form 4562, Depreciation and Amortization*, US Department of the Treasury, Internal Revenue Service, published yearly.

which reduces to:

$$P = 1 - \sum -t\psi \tag{1.11}$$

Ψ is the symbol normally used to denote the present value of \$1 of future depreciation. It can be calculated for any depreciation method, for any time period, and for any return rate. Table 1.5 is a listing of Ψ values for selected depreciation methods.

Table 1.5 The Ψ Factor, the Present Value of $1 of Depreciation

Item no.	Type of depreciation	Tax timing	$\Psi_{i,n}$ with continuous interest
1	Instantaneous	Instantaneous	1
2	Uniform	Continuous	$\dfrac{1-e^{-in}}{in}$
3	Uniform	Periodic, end of year	$\dfrac{1-e^{-in}}{n(e^i-1)}$
4	Flow declining in a straight line to zero; simulated sum-of-the-years-digits	Continuous	$\dfrac{2}{in}\left(1-\dfrac{1-e^{-in}}{in}\right)$
5	Sum-of-the-years-digits	Periodic, end of year	$\dfrac{2}{n(n+1)(e^i-1)}\left(n-\dfrac{1-e^{-in}}{e^i-1}\right)$
6	Declining balance without switch	Periodic, end of year	$\dfrac{F_{DB}}{e^i-(1-F_{DB})}\left[1-\left(\dfrac{1-F_{DB}}{e^i}\right)^n\right]$

Source: K. K. Humphreys, *Jelen's Cost and Optimization Engineering*, 3nd ed., 1991, McGraw-Hill, Inc., New York, p. 99. Reprinted by permission.

Table 1.6 summarizes the unacost and capitalized cost relationships on an after-tax basis.

Example 1.1

Two machines have the following cost comparisons. If money is worth 10% per year, the tax rate is 34%, and sum-of-the-years-digits depreciation is used, which machine is more economical? What is the savings per year?

	A	B
First cost, $	18,000	24,000
Uniform end-of-year expense, $	1000	0
Salvage value, $	500	0
Service life, years	2	3
Life for tax purposes, years	5	5

Solution: For machine A, refer to Table 1.6 and use the unacost expressions for items 1, 2, and 7 in order.

Table 1.6 Summary of Unacost and Capitalized-Cost Relationships—with Tax

Item no.	Item	Present value as a cost for n-year duration	Unacost		Capitalized cost	
			Algebraic expression	Factor expression for sum-of-the-years-digits depreciation	Algebraic expression	Factor expression for sum-of-the-years-digits depreciation
1	C_d Fully depreciable part of an initial cost for n year life, n' years for tax purposes; $C_d = C_i - C_{sal}$	$C_d(1 - t/n_{m'})$	$C_d(1 - t/n_{m'}) \times \dfrac{r(1+r)^n}{(1+r)^n - 1}$	$C_d(1 - tF_{SDP,r,n'})F_{PR,r,n}$	Multiply corresponding algebraic expression for unacost by $\dfrac{1}{r}$	$C_d(1 - tF_{SDP,r,n'})F_{PK,r,n}$
2	R Uniform end-of-year annual cost, depreciated fully at instant incurred	$R(1-t)\dfrac{(1+r)^n - 1}{r(1+r)^n}$	$R(1-t)$	$R(1-t)$		$R(1-t)\dfrac{1}{r}$
3	R_b Uniform beginning-of-year annual cost, fully depreciated at end of year	$R_b\left(1 - \dfrac{t}{1+r}\right)(1+r) \times \dfrac{(1+r)^n - 1}{r(1+r)^n}$	$R_b\left(1 - \dfrac{t}{1+r}\right)(1+r)$	$R_b(1 - tF_{SDP,r,1})(1+r)$		$R_b(1 - tF_{SDP,r,1})F_{PK,r,1}$
4	C_{ex} Irregular cost at end of xth year for article lasting n years, fully	$C_{ex}\dfrac{1-t}{(1+r)^x}$ or $C_{ex}\dfrac{1-t}{(1+r)^x} \times \dfrac{r(1+r)^n}{(1+r)^n - 1}$	$C_{ex}\dfrac{1-t}{(1+r)^x}\times\dfrac{r(1+r)^n}{(1+r)^n-1}$	$C_{ex}\dfrac{1}{(1+r)^x} \times [1 - t(1+r)F_{SDP,r,1}]F_{PR,r,n}$		$C_{ex}\dfrac{1}{(1+r)^x} \times [1 - t(1+r)F_{SDP,r,1}]F_{PK,r,n}$

		depreciated at instant incurred	$\dfrac{C_{ex}}{(1+r)^x} \times [1 - t(1+r)\psi_{r,1}]$			
5	C_{bx}	Irregular cost at beginning of xth year for article lasting n years, fully depreciated at end of year x	$C_{bx}\left(1 - \dfrac{t}{1+r}\right)$ $\times \dfrac{\dfrac{1}{(1+r)^{x-1}}}{\dfrac{r(1+r)^n}{(1+r)^n - 1}}$ or $\dfrac{C_{bx}}{(1+r)^{x-1}}(1 - t\psi_{r,1})$		$C_{bx}\dfrac{1 - tF_{SDP,r,1}F_{PR,r,n}}{(1+r)^{x-1}}$	$C_{bx}\dfrac{1 - tF_{SDP,r,1}F_{PK,r,n}}{(1+r)^{x-1}}$
6	C_{nd}	Nondepreciable first cost such as land or an article that lasts forever	$C_{nd}\dfrac{(1+r)^n - 1}{(1+r)^n}$	$C_{nd}r$	$C_{nd}r$	C_{nd}
7	C_{sal}	Salvage value at end of nth year; treated as a nondepreciable first cost, an expense	$C_{sal}\dfrac{(1+r)^n - 1}{(1+r)^n}$	$C_{sal}r$	$C_{sal}r$	C_{sal}

Source: K. K. Humphreys, *Jelen's Cost and Optimization Engineering*, 3rd ed., 1991, McGraw-Hill, Inc., New York, pp. 74–75. Reprinted by permission.

$$C_d(1 - tF_{\text{SDP},10\%,5})\, F_{\text{PR},10\%,2}$$
$$= (18{,}000 - 500)\,[1 - 0.34\,(0.806)]\,(0.576) = \$7318$$
$$R(1 - t) = 1000(1 - 0.34) \qquad\qquad\qquad = \quad 660$$
$$C_{\text{sal}}(r) \quad = 500(0.10) \qquad\qquad\qquad\qquad\; = \quad\; 50$$
$$R_A = \$8028$$

For machine *B* by item 1,

$$C_d(1 - tF_{\text{SDP},10\%,5})F_{\text{PR},10\%,3}$$
$$= 24{,}000[1 - 0.34\,(0.806)](0.402) = \qquad R_B = \$7004$$

Machine B is more economical with a uniform annual savings after taxes at the end of each year of $(8028 - 7004) = \$1024$. The savings should be reported on a before-tax basis and is $1024/(1 - 0.34) = \$1552$.

The following example is an interesting application. The principles apply when taxes are included, but for simplicity the example omits taxes.

Example 1.2*

A machine costs $20,000 and lasts 6 years with $1000 salvage value at all times. Past records give the following data, where costs are equivalent costs at the beginning of the year and include operating and repair costs.

Year	Costs
1	$3000
2	3500
3	8000
4	15,000
5	5000
6	15,000

If money is worth 10% per year, how long should the machine be kept?

The solution is given in Table 1.7 which shows the unacost for all possible durations of service.

1.5.1 Continuous Interest

Interest has been treated so far as a discrete quantity to be compounded periodically, usually on an annual basis. This is adequate for many purposes, but it is more realistic to calculate interest continuously in conjunction with a business's relatively continuous cash flow. For continuous interest the operator $(1 + i)^{\pm in}$ is

---

*Source: K.K. Humphreys, ed., Jelen's Cost and Optimization Engineering, 3rd ed. Copyright 1991 by McGraw-Hill, Inc. New York. Reprinted by permission.

Table 1.7   Solution to Example 1.2

Unacost basis, 1 yr:

$$\left(20{,}000 + 3000 - \frac{1000}{1.10}\right)F_{PR,10\%,1} = (23{,}000 - 909)(1.10) = \$24{,}300$$

Unacost basis, 2 yr:

$$\left[20{,}000 + 3000 + \frac{3500}{1.10} - \frac{1000}{(1.10)^2}\right]F_{PR,10\%,2} = (26{,}182 - 826)(0.57619) = \$14{,}610$$

Unacost basis, 3 yr:

$$\left[23{,}000 + \frac{3500}{1.10} + \frac{8000}{(1.10)^2} - \frac{1000}{(1.10)^3}\right]F_{PR,10\%,3} = (32{,}794 - 751)(0.40211) = \$12{,}885$$

Unacost basis, 4 yr:

$$\left[23{,}000 + \frac{3500}{1.10} + \frac{8000}{(1.10)^2} + \frac{15{,}000}{(1.10)^3} - \frac{1000}{(1.10)^4}\right]F_{PR,10\%,4}$$
$$= (44{,}064 - 683)(0.31547) = \$13{,}685$$

Unacost basis, 5 yr:

$$\left[23{,}000 + \frac{3500}{1.10} + \frac{8000}{(1.10)^2} + \frac{15{,}000}{(1.10)^3} + \frac{5000}{(1.10)^4} - \frac{1000}{(1.10)^5}\right]F_{PR,10\%,5}$$
$$= (47{,}479 - 621)(0.26380) = \$12{,}361$$

Unacost basis, 6 yr:

$$\left[23{,}000 + \frac{3500}{1.10} + \frac{8000}{(1.10)^2} + \frac{15{,}000}{(1.10)^3} + \frac{5000}{(1.10)^4} + \frac{15{,}000}{(1.10)^5} - \frac{1000}{(1.10)^6}\right]F_{PR,10\%,6}$$
$$= (56{,}796 - 564)(0.22961) = \$12{,}911$$

Minimum unacost is achieved by keeping the machine 5 years.

Source: K. K. Humphreys, *Solutions Manual to Accompany Jelen's Cost and Optimization Engineering*, 3rd ed., 1990, McGraw-Hill, Inc., New York, pp. 153–154. Reprinted by permission.

replaced by $e^{\pm in}$. Thus:

$$S = Pe^{\pm in} \tag{1.12}$$

where e is the Naperian constant (base of natural logarithms) 2.71828 .... As with discrete interest, the sign of the exponent indicates whether time is flowing with or against the calendar. Since interest is now compounded on a continuous basis, the return is somewhat higher than it would be compounded on a periodic basis at the same nominal rate. Frequently with continuous interest (or interest compounded periodically at intervals less than a year) the total return for one year is called the *effective interest rate*, that is, the rate that would produce the same yield if it were compounded annually. Table 1.8 shows the impact of

Table 1.8 Comparison of Compounding Factors

Period	Relationship	For $i = 0.06$	Factor for $i = 0.06$
Annually	$(1 + i)^1$	1.06^1	1.06000
Semiannually	$\left(1 + \dfrac{i}{2}\right)^2$	1.03^2	1.06090
Quarterly	$\left(1 + \dfrac{i}{4}\right)^4$	1.015^4	1.0613635
Monthly	$\left(1 + \dfrac{i}{12}\right)^{12}$	1.005^{12}	1.0616778
Daily	$\left(1 + \dfrac{i}{365}\right)^{365}$	1.00016^{365}	1.0618305
Continuously	e^i	$e^{0.06}$	1.0618365

Source: K. K. Humphreys, *Jelen's Cost and Optimization Engineering*, 3rd ed., 1991, McGraw-Hill, Inc., New York, p. 85. Reprinted by permission.

compounding at the same rate but with different periods and continuously. Table 1.9 shows the relationships between continuous interest and the various methods of assessing cost. Cost comparisons are based on unaflow, R, a uniform annual flow.

1.6 INFLATION

Inflation is an important factor in making cost comparisons. It is designated by d and, like interest rate or rate of return, is expressed in decimal form. Table 1.10 shows the mathematical relationship of various cost items based on capitalized cost. This table should only be used for cases where the rate of return after taxes, r, is greater than the rate of inflation, d. Any other situation is not a practical consideration. It is the after-tax rate of return that establishes the economic viability of any project.

Instead of using the capitalized costs as a common denominator for all possible service lives, items with different service lives can be compared on the basis of unaburden, R_D. *Unaburden* is unacost adjusted for inflation. It bears no relationship to burden or overhead, familiar cost considerations in estimating (see Appendix B). R_D is an equivalent end-of-year cost increased by a factor of $(1 + d)$ to account for inflation. Unaburden can be obtained from capitalized costs by the relationship

$$R_D = (r - d)K \qquad (1.13)$$

Table 1.9 Summary of Relationships for Continuous Interest.

Item no.	Item	Description	Algebraic relationship	Factor relationship
1	P to S	Moves a fixed sum P to another instant of time n years with the calendar	$S = Pe^{in}$	$S = PF_{PS,i,n}$
2	S to P	Moves a fixed sum S to another instant of time n years against the calendar	$P = Se^{-in}$	$P = SF_{SP,\bar{i},n}$
3	$\bar{R}$ to P	Converts a uniform flow $\bar{R}$ for n years to present value at the start of the flow	$P = n\bar{R}\dfrac{1-e^{-in}}{in}$	$P = n\bar{R}F_{RP,\bar{i},n}$
4	R for 1 year to P	Present value of 1 year of uniform flow starting X years hence	$P = \bar{R}e^{-iX}\dfrac{1-e^{i}}{i}$	$P = \bar{R}F_{SP,i,X}F_{RP,\bar{i},1}$
5	P of flow changing at an exponential rate for n years	Present value of $R_x = R_0 e^{\mp gX}$ for n years	$P = n\bar{R}_0\dfrac{1-e^{-(i\mp g)n}}{i\mp g}$	$P = nR_0 F_{RP,(\bar{i}\mp g),n}$
6	P of flow declining in a straight line to zero	Flow goes from $\bar{R}_0$ at zero time to zero in n years; total flow Q is $nR_0/2$	$P = Q\left[\dfrac{2}{in}\left(1-\dfrac{1-e^{-in}}{in}\right)\right]$	$P = QF_{SDP,\bar{i},n}$
7	P to $\bar{R}$	Converts a present value to a uniform flow of n years	$\bar{R} = \dfrac{P}{n}\dfrac{1}{(1-e^{-in})/in}$	$R = \dfrac{P}{n}\dfrac{1}{F_{R\bar{P},\bar{i},n}}$
8	$\bar{R}$ to S	Converts a uniform flow for n years to a future amount n year hence	$S = n\bar{R}e^{in}\dfrac{1-e^{-in}}{in}$	$S = n\bar{R}F_{PS,i,n}F_{RP,\bar{i},n}$
9	S to $\bar{R}$	Converts a future sum S, n years from now, to a uniform flow; sinking-fund payment	$\bar{R} = \dfrac{Se^{-in}}{n}\dfrac{1}{(1-e^{-in})/in}$	$R = \dfrac{S}{n}F_{FS,\bar{i},n}F_{RP,\bar{i},n}$
10	P to K	Converts a present value representing n years to a capitalized cost	$K = P\dfrac{e^{in}}{e^{in}-1}$	
11	K to $\bar{R}$	Converts a capitalized cost to a uniform flow	$\bar{R} = iK$	

Source: K. K. Humphreys, *Jelen's Cost and Optimization Engineering*, 3rd ed., 1991, McGraw-Hill, Inc., New York, pp. 96–97. Reprinted by permission.

Table 1.10 Capitalized-Cost Relationships with Inflation, $r > d$*.

Item no.	Item	Algebraic relationship for capitalized cost	Factor relationship for capitalized cost $K_{r>d}$ sum-of-digits depreciation
1	C_d Depreciable first cost†	$C_d(1 - t\psi_{r_n}) \dfrac{(1+r)^n}{(1+r)^n - (1+d)^n}$	$C_d(1 - tF_{SDP,r,n''}) \dfrac{F_{PS,r,n}}{F_{PS,r,n} - F_{PS,d,n}}$
2	R Uniform end-of-year burden†	$R(1-t)\dfrac{1+d}{r-d}$	$R(1-t)\dfrac{1+d}{r-d}$
3	R_b Uniform beginning-of-year burden†	$R_b(1 - t\psi_{r_1})\dfrac{1+r}{r-d}$	$R_b(1 - tF_{SDP,r,1})\dfrac{1+r}{r-d}$
4	C_{ex} Irregular cost at end of xth year†	$C_{ex}(1-t)(1+d)^x \dfrac{(1+r)^{n-x}}{(1+r)^n - (1+d)^n}$	$C_{ex}(1-t)F_{PS,d,x}\dfrac{F_{SP,r,n-x}}{F_{SP,r,n} - F_{SP,d,n}}$
5	C_{bx} Irregular cost at end of xth year†	$C_{bx}(1 - t\psi_{r_1})(1+d)^{x-1} \dfrac{(1+r)^{n-x+1}}{(1+r)^n - (1+d)^n}$	$C_{bx}(1 - tF_{SDP,r,1})F_{PS,d,x-1}\dfrac{F_{PS,r,n-x+1}}{F_{PS,r,n} - F_{PS,d,n}}$
6	C_{nd} Nondepreciable first cost†	$C_{nd}1$	C_{nd}
7	C_{sal} Salvage value†‡	$C_{sal}\left[1 + t\dfrac{(1+d)^n - 1}{(1+r)^n - (1+d)^n}\right]$	$C_{sal}\left(1 + t\dfrac{F_{PS,d,n} - 1}{F_{PS,r,n} - F_{PS,d,n}}\right)$

*For $d > r$, $K_{d>t} = -K_{r>d}$. For unaburden, $R_d = (r - d)K_{r>d}$ for $r>d$ or $d > r$.
†All costs in terms of present dollar.
‡If excess salvage value due to inflation can be taken as capital gain, use such tax for t in item 7.
Source: K. K. Humphreys, *Jelen's Cost and Optimization Engineering*, 3rd ed., 1991, McGraw-Hill, Inc., New York. p. 139. Material has been added which does not appear in the original source.

Example 1.3

A machine lasts 2 years with negligible salvage value, can be written off in 1 year for tax purposes, and can be purchased new for $10,000. How much can be spent now for a new machine that lasts 10 years, with negligible salvage value, and is written off in 10 years using straight-line depreciation? Money is worth 16% per year after a 34% tax, and the inflation rate is expected to be 10%.

   Solution: In this example, a straight-line calculation is used for illustrative purposes. For United States-based investments, MACRS would actually be used for tax purposes. It should be noted, however, that estimators commonly make estimates and cost comparisons using the straight-line technique for reasons of simplicity. Inherent errors in data and in certain estimating assumptions tend to cancel out the error caused by assuming straight-line depreciation rather than MACRS or sum-of-the-years-digits.

   Unaburden for the 2-year machine from item 1 of Table 1.10 is

$$R_D = (r - d)(C_d)(1 - tF_{SLP,16\%,1}) \frac{F_{PS,16\%,2}}{F_{PS,16\%,2} - F_{PS,10\%,2}}$$

where F_{SLP}, the straight-line present worth factor,

$$F_{SLP} = \frac{(1 + i)^n - 1}{ni(1 + i)^n}$$

Therefore,

$$R_d = (0.16 - 0.10)(10,000)[1 - 0.34(0.862)] \frac{1.346}{1.346 - 1.210}$$

$$= \$4198$$

For the same unaburden for the 10-year machine

$$4198 = (r - d)(C_d)(1 - tF_{SLP,16\%,10}) \frac{F_{PS,16\%,10}}{F_{PS,16\%,10} - F_{PS,10\%,10}}$$

$$= (0.16 - 0.10)(C_d)[1 - 0.34(0.483)] \frac{4.411}{4.411 - 2.594}$$

$$C_d = \$34,484$$

That is, $34,484 could be spent for the 10-year machine. If the inflation rate were zero, it would be permissible to spend only $25,441 for the 10-year machine (calculation not shown). Thus inflation generally adds favor to the longer-lasting article.

1.7 THE LEARNING CURVE

Productivity increases with time. This improvement is commonly associated with improvements in efficiency brought about by increased experience and skill levels. Figure 1.1 shows the direct workhours per pound of airplane against the cumulative number of planes produced for eight types of fighters produced by four manufacturers in World War II. If the same data is plotted on log-log paper, a straight line results as shown in Fig. 1.2. This relationship can be used to generate the equation

$$E_N + KN^s \tag{1.14}$$

where E_N is the effort per unit of production required to produce the Nth unit. K is a constant, derived from the data at hand, that represents the amount of theoretical effort required to produce the first unit. And S is the slope constant, which will always be negative since increasing experience and efficiency leads to reduced effort on a given task. Note that:

$$\frac{E_2}{E_1} = \frac{K(2^s)}{K(1^s)} = 2^s$$

and

$$\frac{E_4}{E_2} = \frac{K(4^s)}{K(2^s)} = 2^s$$

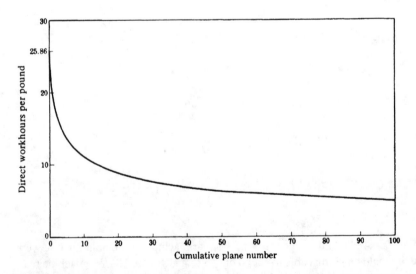

Figure 1.1 Industry average unit curve for century series aircraft. (From K. K. Humphreys, *Jelen's Cost and Optimization Engineering*, 3rd ed., 1991, McGraw-Hill, Inc., New York, p. 212. Reprinted by permission.)

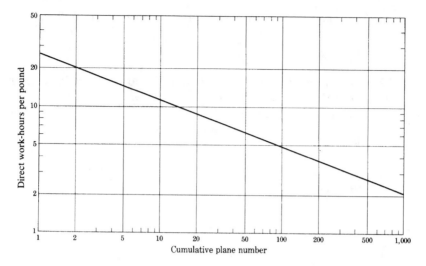

Direct work-hours per pound

Cumulative plane number

Figure 1.2 Log–log plot of industry average unit curve for century series aircraft. (From K. K. Humphreys, *Jelen's Cost and Optimization Engineering*, 3nd ed., 1991, McGraw-Hill, Inc., New York, p. 213. Reprinted by permission.)

and so on. Every time cumulative production is doubled, the effort per unit required is a constant 2^S of what it had been. It is common to express the learning curve as a function of the gain for double the production. Thus a 90% learning curve function means it requires only 90% of the effort to produce the $(2N)$th unit as it did the Nth unit.

If we designate the percentage learning ratio as L_p the relationship between it and the slope is

$$-s = \frac{2 - \log L_P}{\log 2} \tag{1.15}$$

For the total effort required for N units from 1 through N, the cumulative effort, E_T, is

$$E_T = \sum_1^N E_N \tag{1.16}$$

An approximation of this summation is given by

$$E_T = \frac{K}{s+1}[(N+0.5)^{s+1} - 0.5^{s+1}] \tag{1.17}$$

This approximation improves as N increases. If N is very large in comparison to 0.5, (1.17) reduces to

$$E_T = \frac{K}{s+1}N^{s+1} \tag{1.18}$$

The following examples show some practical applications for the learning curve.

Example 1.4*

If 846.2 workhours are required for the third production unit and 783.0 for the fifth unit, find the percentage learning ratio and the workhours required for the second, fourth, tenth, and twentieth units.

Solution: By Eq. (1.14)

$$E_3 = 846.2 = K(3^s)$$
$$E_5 = 783.0 = K(5^s)$$

By division we obtain

$$\frac{846.2}{783.0} = 1.0807 = \left(\frac{3}{5}\right)^s = 0.6^s$$

$$\log 1.0807 = 0.03371 = s \log 0.6 = s(9.77815 - 10) = s(-0.22185)$$

$$s = \frac{0.03371}{-0.22185} = -0.1520$$

By Eq. (1.15)

$$-(-0.1520) = \frac{2 - \log L_P}{0.30103}$$

$$\log L_P = 1.95424$$

$L_P = 90$ percent = percentage learning ratio.
Using the data for $N = 3$ and $S = -0.1520$ in Eq. (1.14),

$$846.2 = K(3^{-0.1520})$$
$$\log 846.2 = \log K - 0.1520 \log 3$$
$$2.9278 = \log K - 0.07252$$
$$\log K = 3.0000$$
$$K = 1,000$$

so that the learning curve function is

$$E_N = 1000N^{-0.1520}$$

Source: K.K. Humphreys, ed., *Jelen's Cost and Optimization Engineering*, 3rd ed. Copyright 1991 by McGraw-Hill, Inc. New York. Reprinted by permission.

The effort required for any unit can now be calculated directly. Thus, for the twentieth unit,

$$E_{20} = 1000(20^{-0.1520})$$
$$\log E_{20} = \log 1000 - 0.1520 \log 20$$
$$= 3.0000 - 0.1520(1.30103)$$
$$= 2.80224$$
$$E_{20} = 634.2$$

A tabulation for other units is:

$$E_1 = 1000.0$$
$$E_2 = 900.0$$
$$E_4 = 810.0$$
$$E_5 = 783.0$$
$$E_{10} = 704.7$$
$$E_{20} = 634.2$$

Note that:

$$\frac{E_2}{E_1} = \frac{E_4}{E_2} = \frac{E_{10}}{E_5} = \frac{E_{20}}{E_{10}} = 0.900$$

Example 1.5*

Every time the production is tripled, the unit workhours required are reduced by 20 percent. Find the percentage learning ratio.

Solution: It will require 80 percent, or 0.80, for the ratio of effort per unit for tripled production.
By Eq. (1.14)

$$\frac{E_{3N}}{E_N} = 0.80 = \frac{K(3N)^s}{KN^s} = 3^s$$

$$s = \frac{\log 0.8}{\log 3} = \frac{9.90309 - 10}{0.47712} = \frac{-0.09691}{0.47712} = -0.20311$$

---

*Source: K.K. Humphreys, ed., *Jelen's Cost and Optimization Engineering*, 3rd ed. Copyright 1991 by McGraw-Hill, Inc. New York. Reprinted by permission.

By Eq. (1.15) for doubled production

$$-(-0.20311) = \frac{2 - \log L_P}{\log 2}$$

$\log L_p = 1.93886$

$L_P = 86.86$ percent = percentage learning ratio

1.8 PROFITABILITY

The six most common criteria for profitability are:

1. Payout time
2. Payout time with interest
3. Return on original investment (ROI)
4. Return on average investment (RAI)
5. Discounted cash flow rate of return (DCFRR)
6. Net present value (NPV)

Consider a project having an income after taxes but before depreciation (cash flow) as follows:

Time, years	Cash flow
0	$-1000
0–1	475
1–2	400
2–3	330
3–4	270
4–5	200

and consider the profitability as measured by these six criteria.

1. *Payout time* is the time to get the investment back. Working capital does not enter payout time calculations since working capital is always available as such. (The project has no stated working capital anyway; working capital is discussed in the next section.) A tabulation gives

Time, end of year	Cumulative cash flow
0	$-1000
1	-525
2	-125
3	+205

The initial investment is returned during the third year; by interpolation, the payout time is 2.38 years.

2. *Payout time with interest* allows for a return on the varying investment. The interest charge is on the fixed investment remaining only; another variant applies an interest charge on working capital. The project leads to the following tabulation with 10% interest:

End of year	Investment for year	Charge on investment for year at 10%	Cash flow	Cash flow after investment charge	Cumulative net cash flow
0			$-1000		$-1000
1	$1000	$100	475	$375	-625
2	625	62.5	400	337.5	-287.5
3	287.5	28.75	330	301.25	+13.75

By interpolation the payout time is 2.95 years, allowing for a 10% return on the varying investment.

3. The *return on original investment* (*ROI*) method regards the investment as fixed. Working capital should be included in the investment figure. In this example working capital is not given and therefore is zero. Using straight-line depreciation and a 5-year project life we get:

Year	Cash Flow	Depreciation	Profit
1	$475	$200	$275
2	400	200	200
3	330	200	130
4	270	200	70
5	200	200	0
			Average = $135

Thus for an original investment of $1000 and no working capital,

$$\text{ROI} = \frac{135}{1000 + 0}(100) = 13.5\%$$

4. The *return on average investment* (*RAI*) is similar to the ROI except that the divisor is the *average outstanding investment* plus working capital. (Again, working capital in this example is considered to be zero.) The average investment for the project with straight-line depreciation is

Year	Investment
1	$1000
2	$1000 - 200 = 800$
3	$800 - 200 = 600$
4	$600 - 200 = 400$
5	$400 - 200 = \underline{200}$
	Average = $600

$$\text{RAI} = \frac{135}{600} 100 = 22.5\%$$

5. The *discounted cash flow rate of return* takes the timing of all cash flows into consideration. This method finds the rate of return that makes the present value of all of the receipts equal to the present value of all of the expenses. For this project we have

$$P = 0 = \frac{-1000}{(1+r)^0} + \frac{475}{(1+r)^1} + \frac{400}{(1+r)^2} + \frac{330}{(1+r)^3} + \frac{270}{(1+r)^4} + \frac{200}{(1+r)^5}$$

By solving the above expression iteratively (i.e., by trial and error) to find a value of r that yields zero, we have

$$r = \text{DCFRR} = 23.9\%$$

6. The *net present value* (*NPV*) is calculated without the need for an iterative solution. Rather, a rate of return is chosen, ordinarily the minimum acceptable rate of return, and the NPV is calculated on that basis. If the minimum acceptable rate of return is 10%,

$$\text{NPV} = \frac{-1000}{(1.10)^0} + \frac{475}{(1.10)^1} + \frac{400}{(1.10)^2} + \frac{330}{(1.10)^3} + \frac{270}{(1.10)^4} + \frac{200}{(1.10)^5}$$

$$\text{NPV} = \$319$$

Table 1.11 gives a simple format for calculating cash flow and the DCFRR. Depreciation is not an out-of-pocket expense, but *must* be considered in calculating the taxable income (tax base); otherwise taxes would be overpaid and net income would drop. Table 1.11 is for a venture that requires $1 million up front and has end-of-year receipts of $600,000, $700,000, and $340,000, for years 1, 2, and 3, respectively. (For simplicity's sake, consider these receipts to be the difference between gross income and expenses, but before taxes and depreciation.) Taxes are 34%, and sum-of-the-years-digits depreciation is used. As shown, for zero net present value for the cash flow, the DCFRR is 22.45%.

Table 1.11 Cash Flow Example

		End year			
	Time 0	1	2	3	Cumulative
1. Total receipts		600,000	700,000	340,000	
2. Depreciation					
$\dfrac{3}{1+2+3}$ 1,000,000		500,000			
$\dfrac{2}{1+2+3}$ 1,000,000			333,333		
$\dfrac{1}{1+2+3}$ 1,000,000				166,667	
3. Taxable income		100,000	366,667	173,333	
4. Tax at 34%		34,000	124,667	58,933	
5. Profit after tax		66,000	242,000	114,400	422,400
Cash flow, item 5 + item 2	−1,000,000	566,000	575,333	281,067	422,400
DCF rate of return, discrete compounding					
Present worth at 20%	−1,000,000	471,648	399,534	162,653	+33,835
Present worth at 25%	−1,000,000	452,800	368,213	143,907	−35,080

For 0 present worth r = 22.45% by interpolation.

1.9 WORKING CAPITAL

Working capital comprises the funds, in addition to fixed capital and startup expenses, needed to get a project started and to meet subsequent obligations. For a manufacturing plant the working capital includes:

- Raw materials inventory
- Supplies for product manufacture
- Work-in-progress inventory (semifinished goods)
- Finished goods inventory
- Accounts receivable
- Accounts payable
- Cash on hand for salaries, wages, etc.

It should be noted that accounts receivable add to the working capital required, while accounts payable reduce the working capital required. In general, working capital is based on the amount of capital needed to cover the above-listed items for a 3-month period. In manufacturing industries, working capital is normally about 10 to 20% of fixed capital investment. In some industries, such as those in sales with a high inventory of finished goods, the figure can be 50% and higher. For a rough estimate, working capital may be taken as 15% of fixed capital plant investment.

1.10 CONCEPTUAL COSTING

Conceptual costing for a project involves estimating the cost of a project with very limited data. It is not required that a flowsheet be available. This method is best adapted to the use of basic cost data from a project using a similar process to the one being considered. For example, basic project cost data for chlorination of benzene might be used for conceptual costing of a project involving similar chlorination of methyl benzene (toluene). Other examples include, but certainly are not limited to, hydrogenation, sulfonation, oxidation, alkylation, and polymerization. In practice, several parameters might have to be included to achieve any degree of accuracy. For example, fixed capital costs for polymerization can vary widely; vinyl chloride polymerizes easily and ethylene polymerizes with difficulty. Here "polymerization" of itself cannot be classified as a definite capital cost. Conceptual costing may be used for the following:

- Ascertaining project profitability in its early development
- Quickly checking bids from contractors and subcontractors
- Estimating costs and profits for potential process licensing
- Quickly assessing inquiries for increased product sales

1.11 FORECASTING

One of the primary tasks of a cost engineer is to make forecasts. Any project analysis requires forecasting at some level, whether it is only for a week or for many years into the future. The cost engineer is not merely asked to estimate the fabrication costs of new facilities. Responsibilities include assessing the benefits of upgrades and changes; making profitability analyses; making accurate schedule forecasts; and breaking down construction, startup, operating, and cumulative costs. The economic success of a project or the feasibility of a course of action all depend upon assumptions the cost engineer has made about coming events.

The forecast starts with an understanding of the present and is based upon experience. Both require sources of data. The best source of data, of course, is an engineer's own company's records and experience; the most common forecast models are based upon in-house history. The next best sources are the professionals and suppliers with other companies or organizations with experience on similar projects. If, however, a project is new, either as a technology or just to the company, published information is usually available if one knows how to go about collecting it. Some additional reading materials and data sources are listed at the end of each chapter. Libraries, especially university and specialty libraries, are lucrative sources of information as is the Internet. For example, unlike many engineering societies, AACE International* maintains a large online technical library of cost engineering publications and data. Staff personnel at libraries and various government agencies can often suggest sources of information that are not immediately obvious. Although this fact in itself seems obvious, staff experts and information specialists are often overlooked by individuals conducting research outside of their familiar sources.

Appendix C lists many Internet sources of data and information. The Internet makes an incredible wealth of information available to project and cost engineers that is invaluable if used with care. It is vital to know that the source is legitimate. The Internet is open to anyone and as such, it has no quality control or veracity check. When using it, use it with caution and common sense.

One must also remember that published information varies widely in quality and applicability. It is risky to base a project decision on print-published or online information without firsthand knowledge of the limitations of the original data. And one must *always* remember that costs and conditions vary widely in different regions and over time. The only real safeguard that a cost engineer has is an understanding of the sources; it is one of the hallmarks of professionalism.

Forecasting has already appeared in the economic relationships that we have discussed. Rates of return and inflation rates have been assumed, and it has been further assumed that these rates remain constant for the length of the project under consideration. These assumptions may not be unreasonable with a short-lived project, but the longer a project stretches, the more tenuous these assumptions become. Often overly conservative estimates are as damaging to a project as is inordinate risk taking (project economics can be "over-engineered" as well as the project itself).

The most common techniques used in forecasting involve trend and regression analysis, which is a mathematical way of saying history matching. Figure 1.3 shows four popular curves used in prediction techniques. Each has its specific areas of applicability.

Computer modeling has become increasingly popular in the field of forecasting. Contributing factors to this increase include the increasing power and affordability of personal or desktop computers and the increasing skill with which business analysts are matching trends with mathematical models. The power extended to forecasters is enormous; parametric analyses that used to require months, and were thus unaffordable in quantity, can now be performed in seconds. All levels of risk can be assessed.

But there is a caveat. As discussed, all forecasts include assumptions. More than one mathematical treatment can be used to analyze the same data.

---

*AACE International, 209 Prairie Avenue, Morgantown, WV 26501, USA. Phone 1-304-296-8444 or 1-800-858-COST. Internet: http://www.aacei.org.

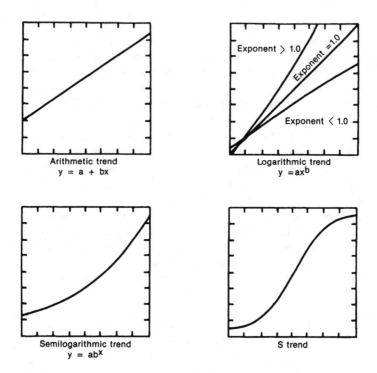

Figure 1.3 Extrapolation of curves for forecasting. The trend: the S-shaped curve plots cumulatively a "normal" of binomial frequency distribution of data. This type of curve is especially valuable in forecasting sales of a specific product, as the life cycles of most products follow this shape. They typically start slowly, pick up momentum as they gain market acceptance, and then taper off gradually as the market becomes saturated and newer products begin replacing them. A symmetrical S-curve plots as a straight line on probability graph paper. (From W. R. Park, *Cost Engineering Analysis*, 1973, John Wiley & Sons, Inc., New York, Reprinted by permission of W. R. Park.)

Unpredictable changes in the economy, such as wars or oil shortages, create changes not addressed by models. In short, computer modeling is only a valuable tool when the modeler has a complete understanding of how the model works. Then, and only then, can the reliability of a model's output be known. Once again, the company's best safeguard is the professionalism of its cost engineer. Unfortunately, there is an increasing tendency for engineers and analysts to use computer software as a "black box" with little personal knowledge of the operational procedures and assumptions made within the "black box" and with unwarranted faith in the output.

1.12 VALUE ENGINEERING

Value and *cost* do not necessarily mean the same thing. Value is total return on the dollar, not just income, but also serviceability, life, efficiency—any consideration that has cost or operational considerations. The following essay, "What is Value?" by D. Lock presents a good overview of the meaning of value.

WHAT IS VALUE?

The concept of value is something which we all apply in our everyday lives when we part with money to acquire an item of furniture or a washing machine or the weekly supply of groceries. Value is not necessarily represented by the lowest-priced product in its class but rather by the one which, for the purposes for which we want it, gives a better combination of ideas, functions, reliability, etc., for its price than the competition.

Value is not an inherent property of a product. It can only be assessed by comparison with other similar products on offer at the same time and under the same conditions. It can be considered in four types:

1. *Exchange value*: that which enables an article to be offered in exchange for money or for another article.
2. *Cost value*: the sum of the materials, labor and other costs required to produce an article or to perform a service.
3. *Use value*: the sum of the properties and qualities possessed by an article which perform a use or provide a service.
4. *Esteem value*: the sum of the features of the article which, beyond its actual use, prompt the decision to buy.

In industry the principal concern must be with use value, the highest value being obtained where the cost is lowest within the limitation that the product must give a satisfactory performance. However, the VA (value analysis) exercise which tries to ignore esteem value or to eliminate it is very often riding for a fall since it will, more than likely, recommend a product which, although near perfection in respect of its performance, is just not saleable—it has lost those features that attract the buyer.

In certain products it is vital to consider the provision of esteem features as one of the essential functions. A water tank hidden in the roof space of a house needs no features other than those concerned with its use but a cooker installed in the kitchen needs many features that have nothing to do with its use or the ladies will never accept it! In summary, good value is the lowest cost reliably to accomplish the essential functions.

UNNECESSARY COST

Value analysis is an unrelenting search for waste, or, put in other words, it is an organized campaign for the elimination of unnecessary cost. Unnecessary cost is defined as any expenditure that can be removed without impairing the quality, reliability, saleability or (if applicable) the maintainability of the subject.

But what allows unnecessary cost to appear? The reasons fall into six main headings that must not be thought to be critical of the design function alone, but are often just as applicable to buying, production, planning and, indeed, to management. They are:

1. *Lack of information*: Is everybody up-to-date in respect to materials, methods, techniques, etc? And, of prime importance, is accurate, detailed and understandable cost information available outside the enclave of accountancy?

2. *Lack of original ideas*: What are people doing with their brains? Using them creatively to find and develop new ideas or reducing them to mere memory banks capable only of responding to the dictates of tradition?

3. *Honest wrong beliefs*: Long-established and honestly-held beliefs can be wrong, even on technical matters, and can lead to unnecessary costs.

4. *Temporary or changed circumstances and time pressures*: A temporary expedient is introduced "to-get-it-into-production." "It's costly but don't worry, old man, it will be put right tomorrow"—and that tomorrow never comes! Or "We shall have to do it this way because of the particular conditions of this order—it will be put right later"—or will it?

5. *Habits and attitudes*: Human beings are creatures of habit and adopt attitudes. The older we get the stronger the habits become and, by consequence, the stronger the built-in resistance-to change. To what extent are personal attitudes towards materials, methods, money and, perhaps, of the greatest importance, men, allowed to influence decisions?

6. *Overdesigning*: This should be considered in its broadest sense and not confined to the designer. Does the specification give the customer exactly what he wants? He may want something much more simple. If so, he is getting poor value.

Source: Reprinted by permission of Gower Press, Teakfield Limited (Aldershot, Hampshire, England) from D. Lock, *Engineers Handbook of Management Techniques*, 1973, 496–97.

Value engineering is a rational or scientific approach to maximizing value. A value engineer tries to "get the biggest bang for the buck." That is, the value engineer attempts to minimize direct costs and eliminate overdesigning while providing serviceability, safety, and favorable returns. It is one area where cost engineering and design work are securely fused into one operation. Because of the complex nature of most projects, it is also a task that requires a team effort, utilizing professionals with many skills and operating at its best with a committee type structure rather than operating under the control of a strong leader. As projects continue to grow in size and complexity, there will continue to be an ever-increasing emphasis on value engineering.

1.13 OPTIMIZATION

Linear and dynamic programming are useful tools in optimization problems and are discussed in some detail in Chapter 8. A chapter on the basics of cost engineering, however, would not be complete without some discussion of optimization techniques. The easiest way to introduce these topics is by use of an example. The following are examples of relatively simple problems in optimization. Theory and situations of greater complexity are discussed in Chapter 8.

1.13.1 Linear Programming

Linear programming is applied linear algebra. A widely used procedure is the simplex method. In this method a feasible solution is used as a starting point, and it is then improved until no further improvements are possible. Linear programming, unlike strict linear algebra, imposes constraints on the problem that prevent unreasonable answers.

Example 1.6

A manufacturer produces two products, A and B. Each product must go through both machine banks 1 and 2; one unit of A requires 2 hours on machine bank 1 and 1 hour on machine bank 2 while one unit of B requires 1 hour on bank 1 and 4 hours on bank 2. Machine bank 1 is limited to 6000 production hours per week, and machine bank 2 is limited to 10,000 hours per week. If the profit on a unit of A is $3.50 and on a unit of B is $5.00, what production schedule will maximize the profit?

Solution: Let x_1 be the number of units of A that are produced and x_2 the units of B. We thus have

$$2x_1 + x_2 \leq 6000 \tag{1.19}$$

$$x_1 + 4x_2 \leq 10,000 \tag{1.20}$$

Since mathematics cannot handle inequalities very well, for our purposes we introduce "slack" variables x_3 and x_4, which are positive in amount, that is, they use up machine time but have no monetary value in profit. Eq. 1.19 and Eq. 1.20 then become equalities and are

$$2x_1 + x_2 + x_3 = 6000 \tag{1.21}$$

and

$$x_1 + 4x_2 + x_4 = 10,000 \tag{1.22}$$

The objective is to maximize

$$z = 3.5x_1 + 5x_2 \qquad (1.23)$$

which represents the weekly profit. W_j will represent the profit realized per unit of production. With three equations and four unknowns, there is no direct solution. But out of an infinite number of solutions, there is only one maximum. Using these three equations, the problem is to set up a feasible mix. A *feasible* mix would be to have only idle time without the actual manufacture of any product $C(x_3)$ or $D(x_4)$. Reducing this to a matrix form, and ascribing all production x_3 to machine bank 1 and x_4 to bank 2, we obtain

	W_j →	3.5	5	0	0		
	Item →	x_1	x_2	x_3	x_4	B_i	
Item	W_j						
x_3	0	2	1	1	0	6,000	$\dfrac{6000}{1}$ = 6000
x_4	0	1	4	0	1	10,000	$\dfrac{10,000}{4}$ = 2500
	Z_j	0	0	0	0	0	↗ smallest
Z_j –	W_j	–3.5	–5	0	0		positive

smallest negative

Rows are designated by i and columns by j. Note that row 1 reduces to Eq. 1.21 and row 2 reduces to Eq. 1.22. The Z_j is the sum of the products of W_j times x_j. Since in our base case there is no production and all x_j thus equal zero, all Z_j are also zero. Notice that the largest negative (or smallest number) in the bottom row is the -5 in the second column. This designates our *pivot column*. By dividing each B by the value in the pivot column, we determine the limiting B value (i.e., the limiting machine bank). Since 10,000/4 is smaller than 6,000/1, B_2 (the variable associated with machine bank 2) is our limiting variable, and row 2 is our pivot row. This means that 4, the element common to both the pivot column and pivot row, is our *pivot element*.

The entire pivot row is now divided by the pivot element and entered into a second matrix. Dividing by 4 yields the row

1/4	1	0	1/4	2500

All other elements in the pivot column must be reduced to zero. This is done by multiplying the other rows by -1 and adding whatever multiples of the pivot row is necessary. Since there is only one other element in the pivot column, a 1, we can merely multiply row 1 by -1 and add row 2 to it:

-2	-1	-1	0	-6000
$1/4$	1	0	$1/4$	2500
$-7/4$	0	-1	$1/4$	-3500

This is now placed in the new matrix as row 1. And since column x_2 was the first pivot column, the x_2 designation and its W value now replace that of x_4 in the left side of the new matrix.

We now have a second matrix that looks like this:

	$W_j \rightarrow$	3.5	5	0	0	
	Item $\rightarrow$	x_1	x_2	x_3	x_4	B_i
Item	W_j					
x_3	0	$-7/4$	0	-1	1/4	$-3,500$
x_2	5	1/4	1	0	1/4	2,500
	Z_j	5/4	5	0	5/4	12,500
$Z_j -$	W_j	$-9/4$	0	0	5/4	

At this point, the entire procedure is repeated. Column 1, containing $-9/4$, becomes the pivot column. Row 1 becomes the pivot row since -3500 divided by $-7/4$ is smaller than 2500 divided by $1/4$. Row one is divided by the pivot element, $-7/4$, and the proper number of multiples of row 1 is subtracted from row 2 to reduce x_{21} to zero. The new matrix thus obtained is:

	$W_j \rightarrow$	3.5	5	0	0	
	Item $\rightarrow$	x_1	x_2	x_3	x_4	B_i
Item	W_j					
x_1	3.5	1	0	4/7	$-1/7$	2,000
x_2	5	0	1	$-1/7$	2/7	2,000
	Z_j	3.5	5	9/7	13/14	17,000
$Z_j -$	W_j	0	0	9/7	13/14	

This third matrix contains no negative number in the last row; this means that there is no possible improvement. Thus profits are maximized at $17,000 by producing 2000 A and 2000 B each week.

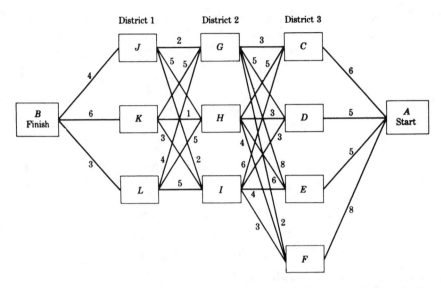

Figure 1.4 Travel costs. (From K. K. Humphreys, *Jelen's Cost and Optimization Engineering*, 3nd ed., 1991, McGraw-Hill, Inc., New York, p. 334. Reprinted by permission.)

Linear programming can become tremendously complex. Here we have used the simplest basic problem: two independent and two dependent variables. There are computer software packages available today that can handle thousands of variables and can maximize *or* minimize profit or expense. The theory, however, remains the same. Chapter 8 presents more detail.

1.13.2 Dynamic Programming

Dynamic programming is useful for finding the optimum policy for a problem involving steps or stages. It can handle nonlinear relationships. The essence of the technique is to reduce the number of policies that must be assessed to find an optimum policy. The following is a typical dynamic programming problem.

Example 1.7*

A sales representative must travel from city A to city B. The territory is divided into three districts, and the representative is required to visit at least one city in each district. Travel costs between cities are indicated in Fig. 1.4. The

---

*This example is taken by permission from K. K. Humphreys, ed. *Jelen's Cost and Optimization Engineering*, 3rd ed., copyright 1991 by McGraw-Hill, Inc., New York. The solution shown here is considerably more detailed than that appearing in the original source.

objective is to visit at least one city in each district at the minimum total travel cost.

Solution: Here the total number of possibilities is only $4 \times 3 \times 3 = 36$, but if there were 10 districts with 10 cities, the number of possibilities would be 10^{10}, a formidable number even for a computer. Dynamic programming is a systematic method that greatly reduces the number of trials in a large problem.

Consider destination city B. There are three routes from district 1 to B, namely JB, KB, and LB. Having arrived at any city in district 1, the sales representative has no remaining choice of route; the costs are

JB = 4	KB = 6	LB = 3

Go back one stage to district 2. All the combinations for going from district 2 to B are

GJB = 6	HJB = 9	IJB = 6
GKB = 11	HKB = 7	JKB = 9
GLB = 7	HLB = 8	ILB = 8

Nothing has been gained in the reduction of trials to this point, but from here on reduction can be realized. The 9 routes just listed can be reduced to the best from G, from H, and from I. They are

GJB = 6	HKB = 7	IJB = 6

Only these 3 (instead of the previous 9) need to be considered in the tabulation for district 3 to city B, which becomes

CGJB = 9	DGJB = 11	EGJB = 14	FGJB = 8
CHKB = 12	DHKB = 10	EHKB = 13	FHKB = 11
CIJB = 12	DIJB = 9	EIJB = 10	FIJB = 9

These 12 routes can be reduced to only 4:

CGJB = 9	DIJB = 9	EIJB = 10	FIBJ = 8

Finally the choice is

A-CGJB = 15
A-DIJB = 14
A-EIJB = 15
A-FGJB = 16

The minimum cost is $14 for route ADIJB.

As with linear programming there are techniques and computer methods for handling problems of extraordinary complexity with the same theoretical foundations. Chapter 8 briefly discusses some of the advantages and limitations of dynamic programming.

1.14 NOMENCLATURE

C	Cost
C_d	Depreciable first cost
C_i	Initial cost for n-year life
C_{sal}	Salvage cost
d	Inflation rate, expressed as decimal
DCFRR	Discounted cash flow rate of return
D_m	Depreciation in mth year, m less than n
D_n	Depreciation for n years
e	Naperian constant (base of natural logarithms), 2.71828...
E_N	Effort required per unit of production for Nth unit
E_T	Cumulative effort required to produce N units
F	Conversion factor, used with subscripts indicating type
i	Interest rate, expressed as decimal
K	Constant used in learning curve calculation
K	Capitalized cost
L_P	Learning percentage ratio
m	Specific year of an n-year operation, m less than n
n	Number of years being considered
N	Unit of production (in learning curve calculations)
NPV	Net present value
P	Present value of money or asset
r	Rate of return, expressed as decimal
R	Uniform end-of-year amount
RAI	Return on average investment
R_b	Uniform beginning-of-year amount
R_D	Equivalent end-of-year burden accounting for inflation
ROI	Return on original investment
s	Slope of learning curve
S	Future value of money or asset
S_n	Future value after n time periods
t	Tax rate, expressed as decimal
Ψ	Factor for present value of \$1 of future depreciation

RECOMMENDED READING

Fallon, C. (1990). *Value Analysis*. 2nd rev. ed. Washington, DC: Lawrence D. Miles Value Foundation.

Hackney, J. W. (1997). Humphreys, K. K., ed. *Control and Management of Capital Projects*. 2nd ed. Morgantown, WV: AACE International.

Humphreys, K. K., ed. (1991). *Jelen's Cost and Optimization Engineering*. 3rd ed. New York: McGraw-Hill Book Company.

Humphreys, K. K., Wellman, P. (1996). *Basic Cost Engineering*. 3rd ed. New York: Marcel Dekker, Inc.

Riggs, J. L., Bedworth, D. D., Randhawa, S. U. (1996). *Engineering Economics*. 4th ed. New York: The McGraw-Hill Companies.

2

Cost Accounting

The cost engineer must be familiar with cost accounting since it is the accountant who keeps the cost records. Moreover the cost engineer should be part of a team in the allocation of overhead and other indirect costs. The first part of this chapter discusses the basic concepts of cost accounting and how they are used on a project. The second part discusses various classifications of accounts. Although the presentation in the first part of this chapter favors work associated with engineering and construction for the oil and chemical industries, the principles are general.

Readers interested in a more detailed discussion of the subject are referred to Humphreys (1991). This book, written under the sponsorship of the AACE International, the Association for the Advancement of Cost Engineering, gives an extensive treatment of the subject.

2.1 PROJECT COST ACCOUNTING

An understanding of the terms *total cost* and *profit* is essential to gain an understanding of cost accounting. For a business to be successful, it must earn a profit. Several terms for sales are employed by accountants, including selling price, contract value, or billable value. They represent the total expected income from the customer.

Considered next are the estimated costs of producing the product, be it a manufactured item or a service. Many firms use the term *total base cost* and commonly have a standard list that defines all the elements that account for the cost of the work. Subtracting the total base cost from the estimated project sales or income gives what is commonly termed *gross profit* or *total overhead and profit*. Another term used is *gross margin*.

Considered also are the costs involved in selling the product or service. The most common terms used are *overhead cost* or *selling and general administration*

43

expenses. These elements of overhead costs are discussed in more detail below. After deduction of the overhead costs there remains the profit or loss before taxes. (Taxes were discussed in Chapter 1).

2.2 ROLE OF COST ACCOUNTING

Cost accounting, the principal subject of this chapter, expands the techniques of financial accounting and is part of managerial accounting, as opposed to historical or audit functions. A general definition of accounting is:

> Accounting is the art of recording, classifying, and summarizing in a significant manner and in terms of money, transactions and events which are, in part at least, of a financial character, and interpreting the results thereof (Grady, 1965).

All accounting systems include three basic steps: recording, classifying, and summarizing economic data, all three in terms of money. Classification of accounts is based on a listing called the *chart of accounts*, which plays a major role in all accounting systems. It is employed extensively in computer applications in the form of management information systems. A good chart of accounts structure provides the following:

- A standard method by which a business prepares cost estimates in a consistent manner
- A means for recording and classifying cost to permit direct comparison with estimates and budgets
- Facilities for creation of cost centers such as department or sections
- Means for dividing cost centers into smaller segments for ease of control and for obtaining unit return cost data
- The opportunity for cost engineers to follow the chart of accounts in trending the project costs and in preparing cost forecasts and management cost reports.

Cost engineers should be completely familiar with their company's chart of accounts. Table 2.1 is an illustration of a general chart of accounts.

2.2.1 Accounting Terms

Before we can consider various classifications of cost, some cost definitions and a few additional accounting terms need to be examined briefly. The definition for "cost" is:

> *... the amount measured in money, cash expended or liability incurred, in consideration of goods and/or services received ...* (AACE, 2003).

The most commonly used terms are listed in Table 2.2; these are all defined either in the text of this chapter or as noted in Appendix B.

Table 2.1 Pro Forma Chart of Accounts

Balance sheet accounts	
2000 Assets	3000 Liabilities
2100 Cash	3100 Accounts payable
2200 Accounts receivable	3200 Notes payable
2300 Notes receivable	3300 Taxes payable
2400 Inventory—materials and supplies	3400 Accrued liabilities
2500 Inventory—finished products	3500 Reserve accounts
2600 Work-in-progress	
2700 Equipment	4000 Equity
2800 Buildings and fixtures	4100 Capital stock issued and outstanding
2900 Land	4200 Retained earnings

Profit-and-loss accounts	
5000 Revenues	6300 Heat, light, and power
5100 Sales of finished goods	6400 Communications expense
5200 Other revenues	6500 Reproduction expense
	6600 Insurance
6000 Expenses	6700 Taxes
6100 Cost of goods sold	6800 Depreciation
6200 Salaries and wages	6900 Interest expense

Cost classification accounts (work-in-progress)	
7000 Construction work-in-progress	7600 Electrical systems
7100 Site preparation	7700 Piping systems
7200 Concrete work	
7300 Structural steel	8000 Manufactured goods-in-progress
7400 Heavy equipment	8200 Direct labor
7500 Buildings	8300 Overhead

These are summary-level accounts, sometimes called *control accounts*. More detail can be provided by using subaccounts as required.
Source: K. K. Humphreys, *Jelen's Cost and Optimization Engineering*, 3rd ed., 1991, McGraw-Hill, Inc., New York, p. 535. Reprinted by permission.

Accounting uses a double-entry system for every transaction with valuation always in terms of money. By way of example, in engineering with no accumulation,

Input = Output

and similarly in accounting

Credit = Debit

The terms *debit* and *credit* are conventional and have no particular significance as such. The distinction between the terms can always be resolved by reducing a

Table 2.2   Accounting Terms

Debit	By-product	Indirect costs[a]
Credit	Process costs	Cost of sales
Assets	Prepaid costs	Variable costs[a]
Liabilities	Standard costs	Burden rate[b]
Manufacturing cost[a]	Accounts payable[a]	FIFO[a]
Cost of goods sold	Fixed costs[a]	LIFO[a]
Joint costs	Direct costs[a]	Gross margin
Unit cost	Accounts receivable[a]	Inventory[a]

[a]*AACE Standard* definition in Appendix B.
[b]See "Burden" in Appendix B.

transaction to a transfer of money between accounts with the accountant acting as an intermediate who transfers the value from one account to another. Suppose that a company pays a supplier $500 cash that is owed to the supplier. The account for the company records a balance transaction as follows:

Supplier	Debit	$500
Cash	Credit	500

Here the supplier's account is a debit since $500 was put into the supplier's account. Similarly the cash account is a credit since $500 was obtained from the cash account.

Two basic classes of accounts must be distinguished:

Real accounts: Real accounts are allowed to accumulate indefinitely and are not closed out at the end of the year.

Revenue and expense, or nominal accounts: These are cleared into capital at the end of the year.

Real accounts are of two types. If they are owned by the business, they are called *assets*. Conversely those owed by the business are called *liabilities*. In accordance with the double-entry equation,

Debits = credits

Assets = liabilities and ownership

where ownership represents the owner's equity in the business.

The meaning of terms used in cost accounting varies from company to company. Hence it is necessary to know exact meaning of terms as used by the cost engineer's company as well as the general meaning. For example, to some accountants the term *manufacturing expense* does not include factory overhead since factory overhead is part of product cost and will funnel into the expense stream only when the product costs are released as cost of goods sold. *Cost of goods sold* is a widely used term that is somewhat misleading in connection with the meaning of cost (see next section). Cost of goods sold is an expense because it is an expired cost and is as much an expense as are sales commissions. Cost of goods sold is often called *cost of sales.*

The distinction between direct costs and indirect costs can cause confusion. The topic is discussed more fully below.

2.3 CLASSIFICATION OF COSTS

Costs can be classified as *unexpired* and *expired*, as summarized in Table 2.3. Unexpired costs (assets) are those which are applicable to the production of future revenues. Expired costs are those which are not applicable to the production of future revenues and for that reason are treated as deductions from current revenues or are charged against retained earnings. Some examples are:

1. *Equipment and machinery.* This could be automobiles, trucks, desks, computers, or machines in a machine shop. All are assigned an estimated useful life, and they are written off or depreciated over a given period of time. The useful life of any piece of equipment is determined by government regulation or by guidelines set by historical data. The guidelines are flexible and may be changed from time to time by modern technology.

When the writeoff is made, the cost becomes an expired cost. The writeoff or depreciation, depending on what it is for, can be classified as direct, indirect, or overhead cost.

2. *Raw materials inventory.* An accumulation of unused items or components (raw materials) related to producing a product is called an inventory. Items in the inventory may have been acquired at different times and at different prices. In such cases a rule or guideline is required to determine the value at

Table 2.3 Unexpired and Expired Costs

Unexpired costs—future revenues
  1. Equipment and machinery
  2. Inventory and raw materials
  3. Prepaid costs
       Insurance
       Taxes
       Rent
  4. Salaries

Expired costs—current revenues
  1. Depreciation expenses
  2. Raw materials (direct materials)
  3. Write-off of prepaid costs
  4. Salaries

which these units shall be transferred out of inventory and into expired costs. The rule applies to pricing and has nothing to do with which unit is removed physically from the inventory.

A common method of inventory costing is *FIFO*, or first in, first out. The cost of the oldest unit in inventory is used. Under conditions of increasing prices or inflation, lower acquisition prices are matched with higher selling prices, resulting in higher accounting profit figures and higher income taxes. To avoid these effects, a method of evaluation can be used known as *LIFO*, or last in, first out, which matches current costs with current revenues. However the lower-cost items remain in inventory. The use of an average valuation represents an effort to find a compromise between the two methods.

3. *Prepaid costs.* These are costs for which cash has been expended for a service or benefit that extends over more than one production cycle, or 1 year. For example, a 3-year $300 insurance premium may be charged originally to an asset account, Prepaid Insurance. Subsequent accounting for the cost will hinge on (a) the amount applicable to the current period, say $100 for the first year, and (b) the purpose of the insurance coverage.

Insurance on factory machinery is inventorial and is therefore transferred from Prepaid Insurance to an inventory account. Insurance on a sales office is not inventorial and is therefore transferred from Prepaid Insurance to an outright expense account. The same determination has to be made for local taxes and rent.

4. *Salaries.* These can be an unexpired cost. Work on a particular product could be done perhaps months before the final assembly of the items is completed. Also a portion of the product could be fabricated outside and returned for final assembly. The salary is written off to the job or to products in the end. Unexpired costs are found in balance sheets, and expired costs are found in the profit and loss (income) section of the final statements.

Cost data found in conventional financial statements, however, are for the specific purpose of reporting to interested outsiders and are not necessarily useful for other applications. The requirement of selecting the appropriate cost type for individual objectives is important. No general rules are possible because of the wide variety of circumstances.

Table 2.4 provides a summary.

Table 2.4   Classification of Costs

Direct costs
Indirect costs
Overhead costs
Standard costs, budgets, and variances
Joint and transfer costs, and pricing
Costing products and services

2.3.1 Direct Costs

Direct costs are also called *prime costs* and are traceable directly to the product being manufactured or fabricated, such as the fabric in clothing. For an engineering and construction firm one of the prime costs on a project is the design department workhours or salary.

In a manufacturing operation, costs are accumulated through three separate accounts:

1. *Direct material.* Cost of materials, assemblies, and parts which are used for the completion of the project
2. *Direct labor.* Wages of workers who are participating in the completion of the product
3. *Overhead.* The costs of all other factors contributing to the completion of the product (see Appendix B)

Material cost consists of the basic purchase price and all other expenses required to transfer the materials to the purchaser's premises, such as transportation, insurance, and tariff duties.

Labor cost is made up of many different factors. It includes the basic hourly rate for hours worked, overtime pay, social security taxes, vacation pay, holidays, sick leave, and so on.

2.3.2 Indirect Costs

Indirect costs are all the costs of manufacturing that cannot be classified as direct costs because it is either impractical or impossible. Each classification is initially accumulated in a separate account and at the end of the accounting period is allocated to individual benefiting activities, such as a cost center or project. It is a two-step procedure: accumulation and allocation.

Supervisory services, for example, benefit many units of profit and the cost accumulates in a separate account. At the end of the accounting period the cost is allocated to individual products as part of the overhead.

Modern technology tends to increase the share of indirect costs while the share of direct labor cost is declining. This is really the purpose of automation—to replace direct labor cost with the indirect cost of machines.

All businesses are organized by departments or cost centers, but the two are not necessarily synonymous. Costs are accumulated by cost centers, which may or may not coincide with operating departments. Cost centers are located where costs can be measured and recorded as conveniently and accurately as possible. The advantages of having departments and cost centers are:

1. A more accurate selling price can be gained for a product.
2. Cost can be controlled more easily.
3. Long-range forecasting can be more exact.

Cost engineers should maintain a close liaison with the cost accountants and those who design and apply the cost accumulation system. Cost engineers need to make the requirements known so that all work can be accomplished within appropriate cost and practicality limitations.

2.3.3 Overhead Costs

The classification includes all product costs that are not considered prime costs (direct material and direct labor). There is no limit to the possible number of overhead classifications; however, principal groups are:

Indirect materials (also known as *supplies*): materials, such as lubricants, that do not become a part of the finished product.

Indirect labor: the wages and salaries of employees who are not directly connected with the manufacture of a product, such as supervisors, maintenance workers, and internal transportation workers. Frequently the cost of fringe benefits is included in this classification.

Facilities costs: both short-term cost of the current year and the depreciated part of long-term costs for the current year. The former includes building and equipment maintenance, local real estate taxes, and other periodic items; the latter includes buildings and equipment.

Service department costs: for facilities that support production—for example, accounting, laboratories, stores, cafeteria, and first-aid stations.

Overhead is considered one of the most tedious problems of cost accounting. Accumulated overhead costs are allocated in stages. First the overhead costs are allocated to the cost centers and in turn are allocated to job order costs and process costs. With a large number of indirect cost classifications and a large number of cost centers, the computations can be voluminous.

Cost engineers must recognize that the cost data that emerge from these calculations are affected significantly by the measures used for allocation, and they should understand fully the techniques used for the allocation. A basis must be selected to allocate the costs fairly. Consider building maintenance as an example. Measurement can be made on the basis of square feet and allocated on the basis of floor space occupied by each cost center. In turn, cost centers must have a measure to allocate this overhead, which might be labor hours, machine hours, material, and so on.

A simple illustration will show how overhead is applied to an individual job at a cost center with allocation based on labor dollars.

Overhead for cost center A	$50,000
Estimated total direct labor cost	$25,000
Burden rate ($50,000/$25,000)	2
for each direct labor dollar	
Direct labor cost for job 504	$2,000
Applied overhead: $2000 × 2	$4,000

Thus job 504 would be allocated $4000 of the total overhead of $50,000 for cost center A.

Allocating overhead based on actual data has two distinct disadvantages: (1) it is complicated, and (2) the information is not timely. Units made at the beginning of a month cannot be priced out until the end of the month when actual data become known, and use of actual overhead cost is subject to fluctuation from period to period for such items as taxes, which may occur only once a year. A shortcut method, called an *estimated burden rate*, has been devised. Managers predict an amount for the overhead cost for a fixed period, usually a year, and determine the measure or basis for charging a job or product for its appropriate share of the overhead cost. Direct labor has been used most widely for this basis.

A rate can be established by dividing total estimated overhead cost by the estimated quantity of the selected basis, which becomes the burden rate. These are estimated amounts and variances will occur. The variances should be reviewed periodically and adjusted accordingly.

The advantages of the estimated burden rate method are (1) savings of time, and (2) increased possibility of obtaining better data for determination of operating efficiency.

2.3.4 Standard Costs

Standard costs and budgets are not identical. Standard costs usually refer to a unit of production, and budget to a total concept like a department. In a sense the standard is the budget for 1 unit of production. The purposes of standard costs are:

- To build a budget and feedback system
- To aid in management predictions
- To save in bookkeeping cost
- To aid in cash flow forecasting

Standard costs are determined with scientific techniques and objective quantity measurements. The costs developed do not necessarily represent expected performance but rather, desired objectives. Standard costs are merely references to which actual costs are compared. The variances are used in management reports as a valuable tool to highlight areas of good and poor performance.

For example, consider a manufacturer producing 10,000 heavy ashtrays. Standards are developed for direct materials and direct labor. The standards based on historical data are:

4 lb at $10/lb = $40/unit	Direct material standard cost
2 hr at $16/hr = $32/unit	Direct labor standard cost

For a budget of 10,000 units, the budget for materials becomes $400,000 and for labor $320,000.

After a run of 10,000 units, the following data are compiled:

Good units produced	10,000
Direct material costs	$540,000
Pounds of input	50,000
Material price per pound	$10.80
Direct labor costs	$343,200
Hours of input	22,000
Labor price per hour	$15.60

Budget variances are determined as follows (F indicates favorable; U indicates unfavorable):

	Total actual costs	Budget total standard costs	Budget variance
Direct materials	$540,000	$400,000	$140,000 U
Direct labor	343,200	320,000	23,200 U

The variance for materials is $140,000 unfavorable, and for labor it is $23,200 unfavorable. The variances can be traced in more detail. The $140,000 unfavorable variance for material arose as follows:

Difference in quantity at $10 standard (50,000 lbs against 40,000 lbs)	$100,000 U
Difference in price for 50,000 lb ($5.40/lb against $5.00/lb)	40,000 U
Total material variance	$140,000 U

The $23,600 unfavorable variance for labor arose as follows:

Difference in hours at $16 standard (22,000 hrs against 20,000 hrs)	$32,000 U
Difference in labor rate for 22,000 hr ($15,60/hr against $16/hr)	8,800 F
Total labor variance	$23,200 U

Note that the unfavorable total labor variance is the sum of an unfavorable workhour overrun and a favorable labor rate. Without the standard cost, analysis is meaningful only if standards are kept current and only if actual data are recorded accurately. Close coordination and cooperation is required between the accountants and the engineers.

2.3.5 Joint Costs

The term *joint costs* applies to two or more kinds of products that are produced simultaneously and are not identifiable as individual types of products until a certain stage of production, the split-off point, is reached. A product for which there is little or no demand is called a *by-product*. Joint costs are combined cost up to the point of separation. Since joint costs cannot be traced directly to units worked, the apportioning of the costs to various units of production has to be arbitrary. Joint cost distribution is limited to purposes of inventory costing and income measurement.

There are two basic approaches for distributing the cost to various products: (1) physical measures, and (2) net realizable value (relative sales value method).

Example 2.1

Consider the following flowsheet:

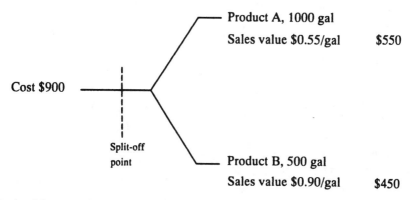

Both of these products become finished goods at the split-off point. Let the $900 cost up to the split-off point be allocated on the basis of the physical measure gallons. For $900 to be spread over 1500 gal total, the cost per gallon is $0.60, and the allocation becomes:

Product	Production	Cost assigned
A	1000 gal	$1000 \times 0.60 = \$600$
B	500 gal	$500 \times 0.60 = 300$
		$900

The income statement becomes:

	A		B		Total
Sales	900 gal	$495	450 gal	$405	$900
Production cost	1000 gal	600	500 gal	300	900
Less inventory	100 gal	60	50 gal	30	90
Cost of sales	900 gal	$540	450 gal	$270	$810
Margin		($45)		$135	$90
Margin percentages:		$\dfrac{45}{495} = -9.1\%$		$\dfrac{135}{405} = 33.3\%$	$\dfrac{90}{900} = 10\%$

There is a loss for A and a profit for B.

Now consider an allocation based on net realizable value.

Product	Net realizable value	Weighting	Cost assigned
A	$550	(550/1000) 900	$495
B	450	(450/1000) 900	405
	$1000		$900

The income statement becomes:

	A	B	Total
Sales	$495	$405	$900
Product cost	495	405	900
Less inventory	49.5	40.5	90
Cost of sales	$445.5	$364.5	$810
Margin	$49.5	$40.5	$90
Margin percentages:	$\dfrac{49.5}{495} = 10\%$	$\dfrac{40.5}{405} = 10\%$	$\dfrac{90}{100} = 10\%$

Now A and B both show the identical margin percentage of 10%. Cost engineers must be aware of the dangers associated with cost allocations for joint products, since the allocation can have a decisive impact on the cost data and on subsequent decision-making.

2.3.6 Job Order and Process Costing

The two basic types of accounting techniques for accumulating production costs are (1) job order costing, and (2) process costing. In job order costing the production cycle and the cost cycle are of equal length. Job order costing is used for specialized production jobs for which costs can be recorded accurately, but for which considerable work is involved. Job orders can be issued for individual customers or for stock items.

Process costing is used for continuous process industries which operate 24 hours per day, such as chemical, steel, and so forth. The production cycle continues without interruption while the cost cycle is cut off for each accounting period to determine the result of operations. Process costs use averages and are less accurate than job order costs, but they are simpler and cheaper to operate. The method is applicable to mass-production industries as well as to process industries.

REFERENCES

AACE Standard Cost Engineering Terminology. (2003). AACE Standard No. 10S-90. Morgantown, WV: AACE International.

Grady, P. (1965). *Inventory of Generally Accepted Principles for Business Enterprises.* AICPA Accounting Research Study 7. New York: American Institute of Certified Public Accountants, p. 2.

Humphreys, K. K. (Ed.) (1991). *Jelen's Cost and Optimization Engineering.* 3rd ed. New York: McGraw-Hill.

RECOMMENDED READING

Adrian, J. J., Adrian, D. J. (1998). *Construction Accounting: Financial, Managerial, Auditing & Tax.* 3rd ed. Champaign, IL: Stipes Publishing Co.

Cokins, G. (2001). Measuring costs across the supply chain. *Cost Eng.* 43(10):25–31; Morgantown, WV: AACE International.

Delaney, P., Epstein, B. J., Nach, R., Budak, S. W. (2003). *Wiley GAAP: Interpretation and Application of Generally Accepted Accounting Principles 2004.* New York: John Wiley & Sons.

MasterFormat: Master List of Titles and Numbers for the Construction Industry. Current ed. Alexandria, VA: Construction Specifications Institute.

Miller, C. A., O'Connell, E. F. (1958). A proposed standard cost code. *AACE Bull.* 1(1):8–11; Morgantown, WV: AACE International.

Palmer, W. J., Coombs, W. E., Smith, M. A. (1995). *Construction Accounting and Financial Management.* 5th ed. New York: McGraw-Hill.

Sillak, G. C. (2002). *Project Code of Accounts.* Recommended Practice No. 20R-98. Morgantown, WV: AACE International.

Sillak, G. C. (2002). *Project Code of Accounts—as Applied in Engineering, Procurement, and Construction for the Process Industries.* Morgantown, WV: AACE International.

3

Cost Estimating

3.1 CAPITAL COST ESTIMATION

The purpose of this chapter is to provide guidelines to the theories and practices of capital cost estimating, primarily for the process industries. Many of the techniques used, however, are generic and can be used widely in other areas of engineering. Additional information is available in the AACE International *Cost Engineers' Notebook*, *AACE Recommended Practices and Standards*, other AACE publications, and in the literature.

3.1.1 Purpose of Estimates

Estimates are the foundation of all cost engineering activity. They are made for several reasons including:

- Feasibility studies
- Selection from alternate designs
- Selection from alternate investments
- Appropriation of funds
- Presentations of bids and tenders

The type of estimate to be made and its accuracy depends upon many factors including the purpose of the estimates, how much is known about the project, and how much time and effort is spent in preparing the estimate. It must be remembered that an estimated project cost is not an exact number. Rather it is an *opinion of probable cost*, not an exact calculation. The accuracy and reliability of an estimate is totally dependent upon how well the project scope is defined and the time and effort expended in preparation of the estimate.

The one thing that is known with 99.9+% certainty about any estimate it that *it is wrong*. Actual cost will rarely, if ever, equal the estimated cost. The final cost of a project will be more or less than the estimate. The difference may be very small, but there will nevertheless be a difference. That fact must always be kept in mind when making project decisions based upon any estimate of costs. It is often said that the only time an estimate is correct is after the project is complete and all of the bills have been paid and totaled.

3.1.2 Fixed Capital and Working Capital

Capital investment is made up of two components:

Total capital = fixed capital + working capital

That is, the total investment to put a project in operation is comprised of:

1. *Fixed capital cost*: a one-time cost for *all* the facilities for the project, including land, design and engineering, equipment, utilities, freight, startup, etc. (see Appendix B).
2. *Working capital*: the funds in addition to fixed capital and land investment that a company must contribute to the project (excluding startup expense) to get the project started and to meet subsequent obligations as they come due. Working capital includes inventories, cash, and accounts receivable minus accounts payable. Characteristically these funds can be converted readily into cash. Working capital is normally assumed to be recovered at the end of the project.

As a rough estimate, working capital for a project will be approximately 10 to 20% of the fixed capital or about 25% of annual operating costs in most manufacturing industries. However working capital requirements can exceed 50% of the fixed capital investment in service industries, industries with high sales costs, and seasonal industries such as construction materials manufacture. As an example, brick is generally manufactured 12 months per year in order to obtain maximum productive use of the kilns used to fire the brick. However in colder climates construction with brick may be possible for only six months or less each year. Thus the brick manufacturers build large inventories during the nonconstruction months, thereby requiring considerably more working capital than might otherwise be needed.

3.1.3 Current Trends in Capital-Cost Estimating

Projects are becoming larger in scale, they are increasingly more international in scope, and competition is keen worldwide. More factors than ever before have to be carefully considered in regard to their impact upon an estimate:

- Technological advances
- Inflation forecasts
- Potential price controls
- Safety and environmental regulations
- Social concerns

- Currency fluctuations
- Import and export considerations
- Taxation and trade barriers
- Escalation
- Inflationary pressures, etc.

3.1.4 Types of Estimates

The American National Standards Institute (ANSI, 1991) defines three types of estimates: order-of-magnitude, budget, and definitive. Definitions of these estimate types were developed by AACE International and are provided in Appendix B.

1. *Order-of-magnitude estimates* have an expected accuracy between +50% and −30%. They are generally based on cost-capacity curves and cost-capacity ratios and do not require any preliminary design work.
2. *Budget estimates* are based on flowsheets, layouts, and preliminary equipment descriptions and specifications and have an accuracy range of +30% to −15%. Design generally must be 5 to 20% complete to permit such an estimate to be performed.
3. *Definitive estimates* require defined engineering data, such as site data, specifications, basic drawings, detailed sketches, and equipment quotations. Design is generally 20 to 100% complete, and estimate accuracy should be within +15% to −5%.

The nonuniform spread of accuracy ranges (e.g., +15% to −5% rather than ±10%) reflects the fact that the vast majority of estimates tend to fall short of actual costs instead of exceeding them.

AACE International has proposed an expansion of the ANSI estimate classifications to five types with expected accuracy levels based upon the amount of project definition available when the estimate is prepared (AACE, 1997). The accuracy of each class of estimate depends on the technological complexity of the project, appropriate reference information, and inclusion of an appropriate contingency determination. In all cases, accuracy ranges could exceed the ranges indicated below in unusual circumstances.

The revised classifications are:

1. *Class 5 estimates*: These estimates are generally based on very limited information. They may be prepared within a very limited amount of time and with little effort expended—sometimes less than one hour. Often little more is known than the proposed plant type, location, and capacity. This class of estimate falls into the ANSI order-of-magnitude classification. The required level of project definition is 2% or less and the expected accuracy is −20% to −50% on the low side and +30% to +100% on the high side.

2. *Class 4 estimates*: Class 4 estimates also are generally prepared based upon limited information and also have fairly wide accuracy ranges. They are typically used for project screening, feasibility determinations, concept evaluation, and preliminary budget approval. Engineering is only 1% to 5% complete and comprises, at a minimum, plant capacity, block schematics, indicated plant layout, process flow diagrams (PFDs) for the main process systems, and preliminary lists of engineered process and utility equipment. Typical accuracy ranges for this class of estimate are -15% to -30% on the low side, and $+20\%$ to $+50\%$ on the high side. This class of estimate falls into the ANSI budget estimate classification.

3. *Class 3 estimates*: These are estimates which form the basis for budget authorization, appropriation, and/or funding. These estimates typically form the initial control estimate against which all actual costs and resources will be monitored. The required level of project definition (i.e., completed engineering) is 10% to 40% and includes at a minimum: process flow diagrams, utility flow diagrams, preliminary piping and instrument diagrams, plot plans, developed layout drawings, and essentially complete engineering process and utility equipment lists. Accuracy ranges for this class of estimate are -10% to -20% on the low side, and $+10\%$ to $+30\%$ on the high side. This class of estimate also falls into the ANSI budget estimate classification.

4. *Class 2 estimates*: This class of estimate falls into the ANSI definitive estimate category. Class 2 estimates are generally prepared to form detailed control baselines against which all project work is monitored in terms of cost and progress control. For contractors, this class of estimate is often used as the "bid" estimate. Typically engineering is 30% to 70% complete and comprises, at a minimum: process flow diagrams, utility flow diagrams, piping and instrument diagrams (P&IDs), heat and material balances, final plot plans, final layout drawings, complete lists of engineered process and utility equipment, single line electrical diagrams, electrical equipment and motor schedules, vendor quotations, detailed project execution plans, resourcing and work force plans, etc. Accuracy ranges are much improved over the prior classes of estimates. On the low side they are -5% to -15%. On the high side, the ranges are $+5\%$ to $+20\%$.

5. *Class 1 estimates*: Also included in the ANSI definitive estimate category, this is the most accurate classification of estimates. Class 1 estimates are generally prepared for discrete parts or sections of the total project rather than for the entire project. The parts of the project estimated at this level of detail are typically used by subcontractors for bids, or by owners for check estimates. The updated estimate is often referred to as the current control estimate and becomes the new baseline for cost/schedule control of the project. This type of estimate is often made to evaluate and/or dispute claims. Typically engineering is 50% to 100% complete and comprises virtually all engineering and design documentation of the projects, and complete project execution and commissioning plans. Typical accuracy ranges are -3% to -10% on the low side and $+3\%$ to $+15\%$ on the high side.

3.1.5 Functions of Capital-Cost Estimates

The various types of capital-cost estimates are made for a variety of reasons, the principal ones being

- Order-of-magnitude estimates
  Feasibility studies
  Selection from alternative designs
  Selection from alternative investments
  Budgeting or construction forecasting
- Budget
  Budgeting or construction forecasting
  Authorization—partial or full funds
- Definitive
  Authorization—full funds
  Check of an authorized project
  Presentation of bids

3.1.6 Cost of Making Estimates

A cost estimate does not come free, and the cost of preparing the estimate rises rapidly with the accuracy required. The expense of an estimate may or may not be calculated to include the cost of detailed engineering required to prepare the estimate. From a practical point of view, the estimate cannot be prepared without adequate engineering work to support the estimate. Thus, whether or not engineering costs are considered to be a part of the cost of making an estimate, the point is moot. For an estimate to be made, engineering must be done to an adequate level to support the estimate. Engineering costs will thus be incurred. Table 3.1 shows the variability of the expense of performing an estimate. This table shows approximately how the estimate preparation cost varies with the size of the project and the level of design.

As shown, for an order-of-magnitude $1 million estimate, estimate cost is about 0.15% of the estimated amount; a budget estimate increases this cost to about 0.6%; a definitive estimate raises the cost to about 2.0%.

Table 3.1 Approximate Cost of Preparing Estimates (Excluding Supporting Engineering Work)

Accuracy range, %	$1,000,000 project	$10,000,000 project	$20,000,000 project
−5/ +15	$20,000	$50,000	$90,000
−15/ +30	6,000	15,000	30,000
−30/ +50	1,500	4,000	8,000

Table 3.2 Approximate Cost of Preparing Estimates (Including Supporting Engineering Work)

Accuracy range, %	$1,000,000 project	$10,000,000 project	$20,000,000 project
−5/+15	$60,000	$130,000	$240,000
−15/+30	35,000	65,000	100,000
−30/+50	14,000	30,000	45,000

Table 3.2 shows the effect of adding the required engineering work to the estimate cost. For the order-of-magnitude $1 million estimate, estimate cost increases to about 1.4% of the estimated amount; the budget estimate increases to about 3.5%; and the definitive estimate raises to about 6.0%.

Clearly, the cost of engineering work must be considered when deciding upon the level of estimate which is required. To perform an estimate at a higher level of accuracy than required is a waste of resources. On the other hand, to attempt to save money by not doing sufficient engineering work leads to estimates which are far less accurate than may be necessary. The estimator must understand the purpose of an estimate and the level of accuracy which is actually needed.

3.1.7   Representation of Cost Data

Cost estimates can be prepared from three sources of data:

1. Similar project costs: costs of similar projects, and costs of project components
2. Proprietary cost data files: historical company costs, and in-house projects
3. Published cost information

Extreme care should be used when using published information because the accuracy level of such data is not known. Sometimes the basis for such data is not even indicated (e.g., whether purchased or installed costs are being presented). Installed cost figures also may or may not include auxiliary equipment, and some costs might be for entire plants while others may be for battery-limit installations only.

Further, unless the published data is dated or includes a cost index value, it is often virtually impossible to correct for inflation since the data was obtained. Do not assume that the publication date reflects the date at which the data was obtained. Publication often takes months or years after an article is first written. Unless the publication is very specific about exactly what is included in the cost figures and about the date of the information, the published data should not be used.

3.1.8   Factored Estimating

The following are typical techniques used to prepare an order-of-magnitude estimate using historical average cost factors. These techniques and the suggested factors must be used with caution and, to the extent possible, be updated based upon actual data for the time and location in question.

Lang Factors

Total plant capital costs (excluding land) can be approximated from the delivered cost of plant equipment using Lang factors as multipliers. The multipliers, depending on the type of plant, are:

- 3.10 for solid process plants
- 3.63 for solid–fluid process plants
- 4.74 for fluid process plants

These factors were originally proposed over 50 years ago by Lang (1948). Despite the passage of time, they continue to be useful for process industry order-of-magnitude estimates.

Hand Factors

Rather than using single multipliers, such as the Lang factors, Hand (1958) suggested using a summation of individual factors multiplied by the delivered cost of different types of equipment. The Hand factors were updated by the AACE International Cost Estimating Committee in 1992 (Hand, 1992). The updated factors are:

- 4.0 for fractionating column shells
- 2.5 for fractionating column trays
- 3.5 for pressure vessels
- 3.5 for heat exchangers
- 2.5 for fired heaters
- 4.0 for pumps
- 3.0 for compressors
- 3.5 for instruments

Wroth (1960) gave a more complete list of factors, as shown in Table 3.3. These factors are also old but are still useful for order-of-magnitude estimates in the process industry.

Plant Cost Estimating by Analytical Procedures

Analytical methods, instead of tables and charts, are particularly useful when using a computer for making cost estimates. Three methods suitable for computer calculation are (1) ratio factor method, (2) Hirsch-Glazier method, and (3) Rudd-Watson method.

Table 3.3   Process Plant Cost Ratio from Individual Equipment

Equipment	Factor[a]
Blender	2.0
Blowers and fans (including motor)	2.5
Centrifuges (process)	2.0
Compressors	
Centrifugals	
Motor-driven (less motor)	2.0
Steam turbine (including turbine)	2.0
Reciprocating	
Steam and gas	2.3
Motor-driven (less motor)	2.3
Ejectors (vacuum units)	2.5
Furnaces (package units)	2.0
Heat exchangers	4.8
Instruments	4.1
Motors, electric	8.5
Pumps	
Centrifugal	
Motor-driven (less motor)	7.0
Steam turbine (including turbine)	6.5
Positive displacement (less motor)	5.0
Reactors—factor as approximate equivalent type of equipment	
Refrigeration (package unit)	2.5
Tanks	
Process	4.1
Storage	3.5
Fabricated and field-erected (50,000 + gal)	2.0
Towers (columns)	4.0

[a]Multiply the purchase cost by a factor to obtain the installed cost, including the cost of site development, buildings, electrical installations, carpentry, painting, contractor's fee and rentals, foundations, structures, piping, installation, engineering, overhead, and supervision.
Source: W. F. Wroth, "Factors in Cost Estimation," *Chemical Engineering*, Vol. 67, October 1960, p. 204.

The *ratio factor method* uses a number of factors added together which are multiplied by the total major equipment cost. Thus

$$Capital\ investment\ for\ plant = (f_1 + f_2 + \cdots)\Sigma E \qquad (3.1)$$

where ΣE is the major process equipment cost, and $f_1, f_2, \ldots$, are cost factors for installation, piping, instrumentation, and so on. The difference between the Hand method and the ratio factor method is in the detail of the factors. A typical summary form that could be used for factored estimates is shown in Figure 3.1, with typical factors for chemical plants.

```
                                    DEPT._____
TITLE____A Chemical Plant_____   STUDY NO._____
     CASE A General Case_____    WORKS Gulf Coast
SCOPE_____ESTIMATOR_____  DATE_____
```

	%	$M
Fabricated equip (process, mechanical, power)____		1000
Misc equipment_____	10	100
Subtotal		1100
Field material/labor/insulation (10 x 15 x 10)____	55	605
Field-erected equipment (eg, storage tanks)		100
Installed equipment		1805
Factored pipe_____	80	1400
Factored instr_____	35	630
Identifiable instr (eg, programmable controller)		50
Factored elec_____	20	360
Identifiable elec (eg, substation, switchgear)		100
MCC/ICR/ECR equip (eg, motor control center)	7	125
Subtotal		4470
Special process items (eg, existing equipment)		(50)
Equip fdns, supports, pltfms_____	12	530
Buildings, structures 96,000 cf @ $8.00/cf		760
Power general/service (identifiable) cooling tower		50
Subtotal		5760
Misc PG&S (eg, outside overhead & u/g lines)____	15	860
Dismantle & rearrange (as required)		
Maintaining production (as required)		
Subtotal (1/m = 40/60)		6620
Working conditions (on labor) existing site - 10%	5	330
Freight/sales tax (on material) 5%/8% Gulf Coast	9	460
Subtotal		7410
Contingencies_____	30	2220
Net total		9630
Auth 3091 mech compl 2093_____	8	770
Direct total		10400
Engineering & project administration_____	35	3650
Project total		14050

Figure 3.1 Factored estimate summary.

The *Hirsch-Glazier* method takes the form

$$I = EA(1 + F_L + F_P + F_M) + B + C \qquad (3.2)$$

where

I = total battery-limit investment
E = indirect cost factor representing contractors' overhead and profit, engineering, supervision, and contingencies; E is normally 1.4
A = total purchased cost f.o.b. less incremental cost for corrosion-resistant alloys
B = installed equipment cost
C = incremental cost of alloys used for corrosion resistance

F_L = cost factor for field labor
F_M = cost factor for miscellaneous items
F_P = cost factor for piping materials

The factors F_L, F_M, and F_P are not simple ratios but are defined by logarithmic relationships.*

The *Rudd-Watson* method takes the form

$$C_{FC} = \phi_1 \phi_2 \phi_3 \Sigma C_{EQ} \qquad (3.3)$$

where

C_{FC} = fixed capital investment
ϕ = components for installation, etc.
C_{EQ} = purchased equipment cost

Note that in the Rudd-Watson method, the factors are multiplied together, whereas in the ratio factor method they are added together.

Plant Component Ratios

Plant cost estimates can be made using factors for separate components. However this method is generally no more accurate than is using an overall factor unless a great deal of information is available. Table 3.4 gives a range of plant direct investment cost for individual components in a typical battery-limit process plant. Indirects should be added for total capital investment.

The figures shown in Table 3.5 also provide some ratios according to major cost categories for a typical chemical plant.

Cost-capacity factors are used to estimate a cost for a new size or capacity from the known cost for a different size or capacity. The relationship has a simple exponential form:

$$C_2 = C_1 \left(\frac{Q_2}{Q_1}\right)^x \qquad (3.4)$$

where

C_2 = desired cost of capacity Q_2
C_1 = known cost of capacity Q_1
x = cost-capacity factor

This equation is often referred to as the *6/10ths Rule* or the *7/10ths Rule* because the average value of the exponent is about 0.6 for equipment and 0.7 for complete plants. However to blindly assume either of these value is dangerous because the

---

*For more details, see K. K. Humphreys, ed., *Jelen's Cost and Optimization Optimization Engineering*, 3rd ed., McGraw-Hill, Inc., New York, 1991, Chapter 14 (or see the original reference, *Chemical Engineering Progress*, Vol. 60, December 1964, pp. 23–24).

Table 3.4 Cost of Components of Typical Chemical Plants

	Percent of direct total	Range, %
Foundations	4	1–8
Equipment	39	26–46
Insulation	4	1–5
Site work	4	1–9
Buildings	10	6–20
Duct work	1	0–3
PWR wiring	8	7–12
Instruments	8	5–10
Piping	22	12–29

factors can, and do, vary within a range of 0.3 to 1.0 and, in some cases over an even wider range.

Rather than assuming a value for the exponent, it is far better to calculate the value if data for various equipment sizes or plant capacities is available or to use published sources of cost capacity factors. Only if nothing is known should the 0.6 value for equipment or the 0.7 value for complete plants be applied, and then only with caution. Further, the use of the equation should be limited to scaleups of no more than 2 to 1 because of structural effects when equipment gets larger or taller. Often discontinuities in the factor occur over a wide range of capacities.

As an example of the use of the equation, assume that the cost for equipment making 100,000 tons/year is $8 million. If 0.6 is a usable value for the exponent, the cost for equipment making 200,000 tons per year would be $12,100,000:

$$C_2 = 8,000,000 \left(\frac{200,000}{100,000}\right)^{0.6} = \$12,100,000$$

Table 3.5 Major Cost Categories in a Typical Chemical Plant

Category	Percent of project range
Engineering design	10–15
Engineering control[a]	2–5
Indirect field costs	15–25
Field labor	30–40
Field material	15–25
Fabricated equipment	25–35

[a]Engineering control includes (1) construction home office expense and drawing reproduction, (2) quality assurance engineering and vendor inspection, and (3) cost control and estimating.

The same procedure can be used for order-of-magnitude estimates for entire plants. Lacking other information, use 0.7 as the cost-capacity factor for entire plants only with extreme caution. Table 3.6 gives some cost-capacity factors for battery-limit plants. Humphreys (1996) has provided an extensive tabulation of cost-capacity factors for both individual equipment items and complete plants.

Table 3.6 Cost Capacity Factors for Process Plants

Product or process	Units of production	Typical plant size	Cost capacity factor
Chemical plants	1000 ton/yr		
Acetic acid		10	0.68
Acetone		100	0.45
Ammonia		100	0.53
Ammonium nitrate		100	0.65
Butanol		50	0.40
Chlorine		50	0.45
Ethylene		50	0.83
Ethylene oxide		50	0.78
Formaldehyde (37%)		10	0.55
Glycol		5	0.75
Methanol		60	0.60
Nitric acid		100	0.60
Polyethylene (high density)		5	0.65
Propylene		10	0.70
Sulfuric acid		100	0.65
Urea		60	0.70
Refinery units	1000 bbl/day		
Alkylation		10	0.60
Coking (delayed)		10	0.38
Coking (fluid)		10	0.42
Cracking		10	0.70
Distillation (atm.)		100	0.90
Distillation (vac.)		100	0.70
Hydrotreating		10	0.65
Reforming		10	0.60
Polymerization		10	0.58

Source: K. K. Humphreys, ed., *Jelen's Cost and Optimization Engineering*, 3rd ed., Copyright 1991 by McGraw-Hill, Inc., New York, Reprinted by permission. Based on M. S. Peters and K. D. Timmerhaus, *Plant Design and Economics for Chemical Engineers*, 3rd ed., McGraw-Hill, Inc., New York, 1980, pp. 184–185.

The following techniques are used for both budget and definitive estimates. Building costs can be estimated from:

- Unit costs (e.g., dollars per square foot)
- Component costs (e.g., floor, wall, roof), or
- Knox volumetric ratio concept that process equipment occupies 3.75% of building volume.

The R. S. Means Company publishes many cost data books and a CD-ROM database on building costs that are quite useful for estimating the cost of buildings, building components, and building systems. Some of their publications are listed in the Recommended Reading list at the end of this chapter.

Cost Indexes

Costs change continuously because of three factors: (1) changing technology, (2) changing availability of materials and labor, and (3) changing value of the monetary unit—that is, inflation. Various cost indexes have been devised to keep up with changing costs. Cost indexes are discussed more fully in Chapter 7. Some frequently used indexes are:

Engineering News-Record Indexes: two indexes—construction and building

Marshall and Swift Equipment Cost Indexes: 47 individual industry indexes plus a composite index representing the installed equipment cost average for these industries

Nelson-Farrar Refinery Construction Index: based on 40% material and 60% labor costs

Chemical Engineering Plant Cost Index: based on four major components:

Equipment machinery and supports	61%
Erection and installation labor	22%
Buildings, material, and labor	7%
Engineering and supervision	10%

Bureau of Labor Statistics indexes: various indexes compiled from national statistics.

Use and Limitations of Cost Indexes

Cost indexes can be used to upgrade a cost for passage of time by a simple proportional relationship. Thus, if a cost was $50,000 when an index was 229, the cost after the index has risen to 245 would be about $53,500:

$$50,000\left(\frac{245}{229}\right) = \$53,500$$

that is

$$C_1 \left(\frac{I_2}{I_1} \right) = C_2 \tag{3.5}$$

where

C_1 = earlier cost
C_2 = new estimate
I_1 = index at earlier cost
I_2 = index at new estimate

Some limitations of cost indexes are:

- They represent composite data, not complete projects.
- They average data.
- They use various base periods for different indexes.
- Their accuracy for periods greater than 5 years is very limited, at best $\pm 10\%$.
- They are highly inaccurate for periods greater than 10 years and, in such cases, should be used for order-of-magnitude estimates only.
- For imported items, indexes do not reflect currency exchange fluctuations and currency reevaluations. These must be considered separately.

For maximum accuracy, index labor and material costs separately if this breakdown is known. This eliminates possible weighting differences between a composite index and a specific project.

Equipment Installation Cost Ratios

A ratio cost factor is merely a factor used to multiply one cost factor by another to get another cost. The principle can be used to obtain installed costs from purchased costs. Thus, if F denotes an equipment installation cost ratio, then

Installed cost = (purchase cost)(F)

Some equipment installation factors are given in Table 3.7. Note that these factors assume that the foundation, hooks-up, etc. for the equipment are in place, and all that is required is to anchor the equipment in position and connect the required, wiring, piping, controls, etc. AACE (1990) and Humphreys (1996) have provided factors for determining the cost of foundations, structures, buildings, insulation, instruments, electrical requirements, piping, painting, and miscellaneous costs associated with equipment installation. These factors reflect the type of material to be processed (solids, gases, or liquids), the process temperature, and process pressure and are applied to the equipment purchase costs.

Table 3.7 Distributive Labor Factors for Setting Equipment

Equipment type	Factor[a]	Equipment type	Factor[a]
Absorber	20	Hammermill	25
Ammonia still	20	Heater	20
Ball mill	30	Heat exchanger	20
Blower	35	Knockout drum	15
Briquetting machine (with mixers)	25	Lime leg	15
Centrifuge	20	Methanator (catalytic)	30
Clarifier	15	Mixer	20
Coke cutter	15	Precipitator	25
Coke drum	15	Regenerator (packed)	20
Condenser	20	Retort	30
Conditioner	20	Rotoclone	25
Cooler	20	Screen	20
Crusher	30	Scrubber (water)	15
Cyclone	20	Settler	15
Decanter	15	Shift Converter	25
Distillation column	30	Splitter	15
Evaporator	20	Storage tank	20
Filter	15	Stripper	20
Fractionator	25	Tank	20
Furnace	30	Vaporizer	20
Gasifier	30	Water scrubber	20

[a]% of bare equipment purchase cost.

Factors to determine the labor cost to set equipment onto prepared foundations/supports includes costs for rigging, alignment, grouting, making equipment ready for operation, etc. The money allowed is to a great extent a matter of judgement. The following general rules are offered as an aid:

1. Equipment such as hoppers, chutes, etc. (no moving parts) require a setting cost of about 10% of the bare equipment purchased cost.
2. Rotary equipment such as compressors, pumps, fans, etc. require a setting cost of about 25% of the bare equipment purchased cost.
3. Machinery such as conveyors, feeders, etc. require a setting cost about 15% of bare equipment purchased cost.

Historical workhour requirements are more desirable than these factors, if available. The factors do not work well for very large equipment. For example, a $750,000 compressor does not require 25% of the bare equipment cost to set same on the foundation and to "run-it-in." The listing above provides approximate factors for specific other types of equipment.

Source: Conducting Technical and Economic Evaluations in the Process and Utility Industries, *AACE Recommended Practices and Standards*, RP No. 16R-90, AACE International, Morgantown, WV, 1990.

Piping costs vary from 1 to 25% of project cost in a process industry facility. Installation costs are three to four times purchase costs for common materials. Order-of-magnitude estimates can be made by use of factors or ratios. Detailed estimates involve quantity take-offs, labor units, and costing.

Intermediate estimates utilize tables and charts for average piping components. The *diameter-inch method* is based on:

(Number of connections) × (pipe diameter) × (labor factor)

An average is 1.5 workhours per diameter-inch. Ratios are also available for materials other than carbon steel.

Several manuals and references are available for estimating piping costs. Table 3.8 presents estimating data from five different sources for installation of 3/4-inch carbon steel piping. It can be readily seen that not all sources cover all situations; indeed none of these five give values for all of the situations listed. Even more importantly, the spread of values for the same situation in different sources is extensive. Figure 3.2 is a band chart of the same data that makes this variation even more noticeable. Because of these variations, estimators and cost engineers need to be highly familiar with their data sources

Table 3.8   Example of Tabular Estimating Data Workhours for Fabrication and Erection of 3/4 inch Carbon Steel Piping

Operation	Source				
	A	B	C	D	E
Butt weld, std wt	1.1	0.8	0.4	—	0.5
Butt weld, X.H.	1.3	0.9	1.2	0.7	0.6
Lin. ft. std. pipe	0.13	0.07	0.2	—	0.19
Lin. ft. X.H. pipe	0.15	0.08	0.25	0.16	0.20
Std. weld ells	2.2	1.6	0.9	2.1	1.3
Std. weld tees	3.3	2.4	1.2	3.0	1.8
X.H. weld ells	2.6	2.7	1.0	2.1	1.5
X.H. weld tees	3.9	2.7	1.5	3.0	2.1
Std. weld reducer	2.0	1.4	0.5	2.1	1.3
X.H. weld reducer	2.2	1.5	0.9	2.1	1.5
150# slip-on flange	1.1	—	0.9	—	0.80
300# slip-on flange	1.2	—	1.0	—	0.90
600# slip-on flange	1.3	—	1.1	—	—
150# weld neck flange	1.1	1.1	0.4	—	0.80
300# weld neck flange	1.2	1.1	1.3	—	0.95
600# weld neck flange	1.3	1.1	1.4	—	—
150# flange bolt-up	—	0.6	1.5	—	—
300# flange bolt-up	—	0.7	1.6	—	—
600# flange bolt-up	—	0.7	—	—	—
Handle 150# valve	—	0.31	1.5	0.8	—
Handle 300# valve	—	—	1.6	—	—
Handle 600# valve	—	—	—	1.1	—

Source: "Data for Estimating Piping Cost," *Cost Engineers' Notebook*, Paper 64–31, June 1964, AACE International, Morgantown, WV. *Cost Engineers' Notebook*, AACE International, Morgantown, WV, June 1964, Paper 64–31.

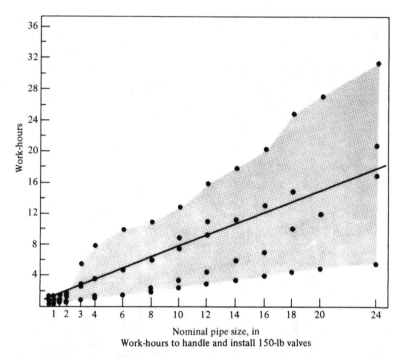

Figure 3.2 Example of a band chart. From K. K. Humphreys, ed., *Jelen's Cost and Optimization Engineering*, 3rd ed., Copyright 1991 by McGraw-Hill, Inc., New York, Reprinted by permission. Originally published in "Data for Estimating Piping Cost," *Cost Engineers' Notebook*, Paper 64–31, June 1964, AACE International, Morgantown, WV.

and with what costs are included in the given cost figures. Blind acceptance of cost references or indexes can lead to disastrous results.

The cost for piping materials, such as straight pipe, fittings, and valves, is generally well established for a host of materials ranging from black iron to stainless steel to plastics. Most companies keep accurate information on the work-hours of labor required to install pipe and fittings for the various sizes and materials in use. Insulation costs usually run 2 to 4% of plant costs. Engineering costs are approximately 10% of a project's cost but will vary with the size and type of job, as is shown in Figure 3.3.

3.1.9 Codes of Account

The capital-cost estimate should be prepared according to an accounting code system. When the estimate is made in such a format, analysis of the project is simplified, and the estimate then becomes the basis of cost reporting and cost control. Table 3.9 gives an example of various types of work categorized by accounting codes that can be used as an estimating tool.

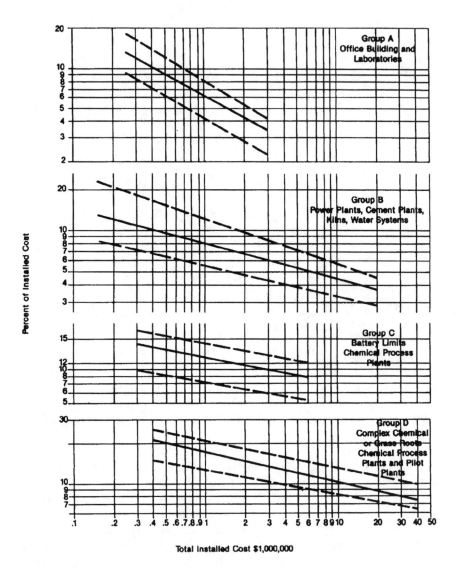

Total Installed Cost $1,000,000

Figure 3.3 Engineering costs. From "Engineering Costs", *Cost Engineers' Notebook*, AACE International, Morgantown, WV, December 1992. Revised and updated from an article by H. Carl Bauman originally published by AACE International, December 1964.

3.1.10 Startup Costs

Startup costs are a capital cost item and are the costs of making the transition from construction to operation. They include costs to make changes so the plant can operate at or near design capacity. Startup costs are generally about 10% of the project cost.

Table 3.9   Typical Cost Estimate Codes of Account

Ac No.	Item and description	Estimated cost				
		Workhours	Labor	Subcontracts	Materials	Total
00.00	Excavation					
10.00	Concrete					
20.00	Structural Steel					
30.00	Buildings					
40.00	Machinery and Equipment					
50.00	Piping					
60.00	Electrical					
70.00	Instruments					
80.00	Painting and Scaffolding					
85.00	Insulation					
	DIRECT FIELD COSTS					
90.00	International Expense					
91.00	Temporary Construction Facilities					
92.00	Constr. Services, Supplies and Expense					
93.00	Field Staff, Subsistence and Expense					
94.00	Craft Benefits, Payroll Burdens and Insur.					
95.10	Equipment Rental					
95.50	Small Tools					
99.40	Field Staff Overhead Costs					
	INDIRECT FIELD COSTS					
	TOTAL FIELD COSTS					
96.00	Home Office Construction					
	Project Engineering					
	Process Engineering					
	Design					
	Purchasing					
	Business Services					
97.00	Office Expense					
98.00	Office Payroll Burdens					
99.50	Office Overhead Costs					
	TOTAL OFFICE COSTS					
	TOTAL FIELD AND OFFICE COSTS					
99.60	Fee					
99.30	Sales Tax					
99.10	Escalation					
99.20	Contingency					
	TOTAL					

3.1.11 Evaluation of the Estimate

After material take-offs and the various estimating techniques previously
described have been applied, several other factors must be considered in order
to complete the total capital investment estimate. Contingencies should be
applied as required to cover undeveloped design, technological unknowns, and
estimate omissions. It is very important to understand that contingency is not a
"slush fund" to cover scope changes. It is an allowance to cover costs that are
known from experience to occur but which can not be determined at the time
the estimate is prepared. As shown in Appendix B, AACE International
defines contingency as:

> An amount added to an estimate to allow for items, conditions, or events
> for which the state, occurrence, and/or effect is uncertain and that
> experience shows will likely result, in aggregate, in additional costs.
> Typically estimated using statistical analysis or judgment based on
> past asset or project experience. Contingency usually excludes:
> (1) major scope changes such as changes in end product specification,
> capacities, building sizes, and location of the asset or project (see man-
> agement reserve), (2) extraordinary events such as major strikes and
> natural disasters, (3) management reserves, and (4) escalation and
> currency effects. Some of the items, conditions, or events for which
> the state, occurrence, and/or effect is uncertain include, but are not
> limited to, planning and estimating errors and omissions, minor price
> fluctuations (other than general escalation), design developments and
> changes within the scope, and variations in market and environmental
> conditions. Contingency is generally included in most estimates, and
> is expected to be expended to some extent.

A more succinct definition of contingency is that of the Association of Cost
Engineers (UK):

> Budgetary provision for unforseen occurrences within a defined project
> scope

A risk analysis should be prepared to determine what level of contingency
is required based on the owner's expectations and requirements. A properly done
risk analysis will identify the amount of contingency required to be added to the
estimate to reduce the probability of a cost overrun to whatever level is deemed
appropriate by the owner. Risk analysis techniques and contingency determination are discussed in some detail in Chapter 9.

Finally, escalation rates should be calculated and applied according to labor
and material categories for the projected expenditures that will occur in each year
of the project schedule.

When the estimate has been completed, the estimate accuracy should be
verified by an independent check. Validation should be made for major parts

of the project and for the overall project. The validation begins by comparing major items (such as design cost, construction management, labor, and material estimated project level percentages) to normal ranges for the company or industry. Next, major project components such as buildings and piping should be compared to similar projects and processes to ensure that the estimate is in line with actual costs. The final validation step should be to use a similar completed project as a basis, and after accounting for quantitative differences between the project and estimate, determine if this new number is close to the estimated number. These validations help explain how the proposed project is different from or similar to other projects and indicate where a closer evaluation should be made. After these validations are successfully made, the estimate may be judged to be reasonable.

3.2 OPERATING COST ESTIMATION

3.2.1 Definitions

Operating Cost (or manufacturing cost): the expenses incurred during the normal operation of a facility, or component, including labor, materials, utilities, and other related costs.

Overhead: a cost or expense inherent in the performing of an operation, i.e., engineering, construction, operating or manufacturing, which cannot be charged to or identified with a part of the work, product or asset and, therefore, must be allocated on some arbitrary base believed to be equitable, or handled as a business expense independent of the volume of production. Plant overhead is also called factory expense.

Direct Costs: the portion of the operating costs that is generally assignable to a specific product or process area.

Indirect Costs: costs not directly assignable to the end product or process, such as overhead and general purpose labor, or costs of outside operations, such as transportation and distribution.

3.2.2 Purposes of Operating Cost Estimates

Operating cost estimates can be important from several standpoints. They are necessary in order to determine the potential profitability of a product or process and are very useful for screening alternative project possibilities. Frequently they can act as a guide to pinpoint potential areas in which to conduct research and to evaluate the commercial viability of research results. And they provide a tool for sensitivity analysis for individual components. It is very important to remember to perform operating cost estimates at both full and reduced capacities for the operating system in question. It is not uncommon for a system to be highly efficient at full capacity, but very inefficient when operating at less than design capacity. The important question to be answered is what *range* of operations can a plant operate over and still make a profit. Such costs can be calculated either as stand-alone estimates or as incremental costs for specified projects. Depending upon the situation, these different viewpoints could result in different decisions.

Commonly, three differing bases for comparison are used, although others may be used for special applications. The most common three are:

- Daily cost
- Cost per unit-of-production
- Annual cost

For some manufacturing considerations, a cost-per-unit analysis can be valuable. When making comparisons, it is very important to make sure each iteration or calculation is made using the same techniques and assumptions. Otherwise, comparisons may not be valid. All other factors being equal, making comparisons using an annualized basis instead of on shorter time increments has a number of advantages:

- Seasonal variations are evened out.
- On-stream time factors are considered.
- More accurate analysis of less-than-full capacity situations is obtained.
- Infrequently occurring large expenses are factored in (scheduled maintenance, vacation shutdowns, catalyst changes, etc.).
- The output is in a form easily used in standard profitability analysis.
- The annual basis is readily convertible to the other bases, daily cost and unit-of-production, yielding mean annual figures rather than a potentially high or low figure for an arbitrarily selected time of year.

3.2.3 Raw Materials and Utilities

Raw materials and/or utilities are frequently the largest operating cost being considered in an estimation. The estimations involving raw materials are not always straightforward, as the following list of considerations shows.

- Raw materials frequently are the largest operating cost.
- By-products and scrap may be a debit or credit.
- Prices may be obtained either from suppliers or from published data.
- Raw materials costs may vary significantly depending upon the quality required (concentration, acceptable impurity levels, etc.).
- Raw materials costs entail quantity discounts in many cases leading to a trade-off of lower prices for purchase of large quantities vs. storage and inventory cost for raw materials that cannot be used immediately.
- Raw materials costs vary significantly depending upon the mode of purchase and transport (bulk quantities, truck lots, rail car lots, pipeline delivery, bags, boxes, etc).
- Supplies and catalysts are sometimes considered to be raw materials.

- Fuels may be either raw materials or utilities.
- Freight and handling costs must be included in pricing raw materials.

As this list illustrates, accepted local practice can have a large impact on the operating cost estimate.

The cost of utilities can vary widely with the location, with the size of the service required, with the national and local economy, and even with the season. Utilities must be examined closely on a local basis, especially if they form a major part of the operating costs. Generally utility pricing is regulated and the approved tariffs are readily available from the utility company or the cognizant regulatory agency. However, if the utilities are internally generated within a process, a separate estimate of the utility cost must be performed.

3.2.4 Operating Labor

The best estimates of operating labor are based on a complete staffing table. This table should indicate the following:

- The particular craft or skill required in each operation
- Labor rates for the various types of operations
- Supervision required for each process step
- Maintenance personnel required
- Overhead personnel required

In the absence of a staffing table, an order-of-magnitude approximation of labor requirements can be made using the Wessell equation. The Wessell equation is

$$\frac{\text{Operating workhours}}{\text{Tons of product}} = T \left[\frac{\text{number of process steps}}{(\text{capacity, tons/day})^{0.76}} \right] \qquad (3.6)$$

where

$T = 23$ batch operations with maximum labor
$T = 17$ for operations with average labor requirements
$T = 10$ for well-instrumented continuous process operations

Supervision costs may vary from 10% to 25% of operating labor cost. On average, an allowance of 15 to 20% of operating labor cost is usually satisfactory for early-on estimates.

3.2.5 Maintenance

Few published data on maintenance data are available. Maintenance cost can vary from 1 or 2% to over 15% of the project capital cost per year. For simple plants with relatively mild, noncorrosive conditions, an allowance of 3 to 5% should be adequate. For complex plants and severe corrosive conditions, this factor can be 10 to 12% or even higher.

Maintenance costs consist of labor and material components with the total maintenance cost divided 50 to 60% labor and 40 to 50% material. For most early-on estimates, it is sufficient to assume a 50 : 50 split.

Maintenance varies with rate of operation, but not linearly. When operating at less than 100% of capacity, maintenance costs generally increase per unit of production as follows:

% of capacity	Maintenance cost as a percentage of cost at 100% of capacity
100	100
75	85
50	75
0	30

Maintenance costs increase with equipment or system age, but the average values are normally used in estimates.

3.2.6   Indirect Payroll Cost

Indirect payroll cost includes workers' compensation, pensions, group insurance, paid vacations and holidays, social security, unemployment taxes, fringe benefits, and so on. It is generally based on labor cost, usually at about 30 to 45% of the labor cost in the United States. In other countries, this percentage may vary greatly and should be verified locally.

3.2.7   Operating Supplies

Operating supplies include lubricants, instrument charts, brooms, and so on, and may be generally assumed to be 6% of the operating labor or as 0.5% to 1% of the capital investment per year. For highly automated, complex operations these costs can increase substantially as a percentage of labor costs.

3.2.8   Laboratory and Other Service Costs

Laboratory and other service costs can be based on one or more of the following:

1. Experience
2. Workhours required
3. A percentage of operating labor cost (3 to 10% is common, but it may be as high as 20%).

3.2.9   Royalties and Rentals

Royalties and rentals are generally an operating expense but may be part of capital investment. Single-sum payments for royalties, rental, or license payments are properly considered as capital investment items, whereas payments in proportion to production or fixed payments per annum are treated as direct operating costs. Royalty payments may range between 1% and 5% of product sales price.

3.2.10 Indirect Costs

Factory overhead is the indirect cost of operating a plant. It is dependent upon both investment and labor. Some data on the allocation for new plants follow (Black, 1991). The figures represent what percentage of the investment and labor, respectively, can be used for estimating the factory overhead.

	Investment factor, % per year	Labor factor, per year
Heavy chemical plants, large capacity	1.5	45
Power plants	1.8	75
Electrochemical plants	2.5	45
Cement plants	3.0	50
Heavy chemical plants, small capacity	4.0	45

Factory overhead may also be called *general works* expense, but it is *not general and administrative expense* (G&A). G&A covers plant administrative costs, accounting, auditing, research and development, marketing and sales expense, etc).

3.2.11 Distribution Costs

Distribution costs include handling and transportation costs. These costs vary with the types of containers and with methods of shipment. Cost without distribution cost is called *bulk cost.*

3.2.12 Avoidance of Nuisances

Nuisances include waste disposal and pollution control costs. Each case must be calculated individually. The topic is becoming increasingly more expensive and includes such items as product liability. These cost are mounting rapidly.

3.2.13 Contingencies

As is true for the capital cost estimate, an operating cost estimate should include a contingency allowance to account for those costs that cannot readily be determined or are too small to be readily determined or defined, but may be significant in the aggregate. The contingency allowance applies to both direct and indirect costs and ranges from 1% to 5% or more in some cases.

Hackney (1971) has suggested the following guidelines for operating cost contingency:

1. Installations similar to those currently used by the company and for which standard costs are available: 1%
2. Installations common to the industry for which reliable data are available: 2%
3. Novel installations that have been completely developed and tested: 3%
4. Novel installations that are in the development stage: 5%

3.2.14 Shortcut Methods

Table 3.10 will help in making preliminary operating cost estimates.

Table 3.10 Preliminary Operating Cost Estimate

A. Direct production cost
    1. Materials
        a. Raw materials—estimate from quotations or price lists
        b. By-product and scrap credit—estimate from price lists
    2. Utilities—from regulatory tariffs
    3. Labor—from manning tables, literature, or similar operations
    4. Supervision—15 to 20% of labor
    5. Payroll charges—30 to 45% of labor plus supervision plus maintenance labor
    6. Maintenance—3 to 5% of investment per year for mild conditions; for severe corrosive conditions, 10 to 12%. Maintenance costs are approximately 50% labor, 50% materials.
    7. Operating supplies—0.5 to 1.0% of investment per year
    8. Laboratory—3 to 20% of labor per year depending upon complexity; 5 to 10% in average situations.
    9. Waste disposal—from literature, similar operations, or separate estimate
    10. Royalties—1 to 5% of sales
    11. Contingencies—1 to 5% of operating costs
B. Indirect costs
    1. Depreciation—assume 10-year straight-line
    2. Real estate taxes—1 to 2% of investment per year
    3. Insurance—0.5 to 1.0% of investment per year
    4. General plant overhead—40 to 60% of labor, supervision, and maintenance
C. Distribution costs
    1. Packaging—estimate from container costs
    2. Shipping—from carriers

Table 3.11 Components of Total Product Cost

I. Operating cost or manufacturing cost
    A. Direct production costs
        1. Materials
            a. Raw materials
            b. Processing
            c. By-product and scrap credit
            d. Utilities
            e. Maintenance materials
            f. Operating supplies
            g. Royalties and rentals
        2. Labor
            a. Direct operating labor
            b. Operating supervision
            c. Direct maintenance labor
            d. Maintenance supervision
            e. Payroll burden on all labor charges
                i. Social security tax
                ii. Medicare tax
                iii. Workers' compensation coverage
                iv. Contributions to pensions, life insurance, hospitalization, and dental plans
                v. Vacations, holidays, sick leave, overtime premium
                vi. Company contribution to profit sharing
    B. Indirect production costs
        1. Plant overhead or burden
            a. Administration
            b. Indirect labor
                i. Laboratory
                ii. Technical service and engineering
                iii. Shops and repair facilities
                iv. Shipping department
            c. Purchasing, receiving, and warehousing
            d. Personnel and industrial relations
            e. Inspection, safety, and fire protection
            f. Automotive and rail switching
            g. Accounting, clerical, and stenographic
            h. Communications—telephone, mail, and teletype
            i. Plant custodial and protective
            j. Plant hospital and dispensary
            k. Cafeteria and clubrooms
            l. Recreational activities
            m. Local contributions and memberships
            n. Taxes on property and operating licenses
            o. Insurance—property, liability

(continued)

Table 3.11 *Continued*

---

    p. Nuisance elimination—waste disposal and pollution control
  2. Depreciation
 C. Contingencies
 D. Distribution costs
  1. Containers and packages
  2. Freight
  3. Operation of terminals and warehouses
    a. Wages and salaries—plus payroll burden
    b. Operating materials and utilities
    c. Rental or depreciation
II. General expense
 A. Marketing or sales costs
  1. Direct
    a. Salespersons' salaries and commissions
    b. Advertising and promotional literature
    c. Technical sales service
    d. Samples and displays
  2. Indirect
    a. Sales and supervision
    b. Travel and entertainment
    c. Market research and sales analysis
    d. District office expenses
 B. Administrative expense
  1. Salaries and expenses of officers and staff
  2. General accounting, clerical, and auditing
  3. Central engineering and technical
  4. Legal and patent
    a. Inside company
    b. Outside company
    c. Payment and collection of royalties
  5. Research and development
    a. Own operations
    b. Sponsored, consultant, and contract work
  6. Contributions and dues to associations
  7. Public relations
  8. Financial
    a. Debt management
    b. Maintenance of working capital
    c. Credit functions
  9. Communications and traffic management
  10. Central purchasing activities
  11. Taxes and insurance

Source: K. K. Humphreys, ed., *Jelen's Cost and Optimization Engineering*, 3rd ed., Copyright 1991 by McGraw-Hill, Inc., New York, Reprinted by permission. Adapted from R. H. Perry et al., Chemical Engineers' Handbook, 5th ed., McGraw-Hill, Inc., New York, 1973, pp. 25–27.

3.2.15 Components of Total Product Cost

Table 3.11 illustrates the connection between the component cost items that enter the total product cost. There are variations from company to company, but this tabulation is still a good guide.

3.3 SUMMARY

The basic cost estimating procedure is as follows:

1. Scope the job or problem.
2. Establish the estimate format; use a code of accounts.
3. Prepare your estimate area by area and category by category.
4. Check your work; verify your data.
5. Review and adjust.
6. Finalize.
7. Complete the documentation.
8. File.

In general,

- Understand the *overall* picture.
- Document well.
- Analyze your figures.
- Tailor your estimate to the job.
- Plan ahead.

REFERENCES

AACE International Recommended Practice No. 16R-90. (1990). *Conducting Technical and Economic Evaluations in the Process and Utility Industries*. AACE International, Morgantown, WV.

AACE International Recommended Practice No. 18R-97. (1997). *Cost Estimate Classification System—as Applied in Engineering, Procurement, and Construction for the Process Industries*. AACE International, Morgantown, WV.

ANSI Z94.0-1989. (1991). *American National Standard Industrial Engineering Terminology*. rev. ed. McGraw-Hill Inc., New York and Institute of Industrial Engineers: Norcross, GA.

Black, J. H. (1991). Operating cost estimation. In: Humphreys, K. K., ed. *Jelen's Cost and Optimization Engineering*. Chap. 15. New York: McGraw-Hill.

Hackney, J. W. (1971). Estimate production costs quickly. *Chem. Eng.* 183.

Hand, W. E. (1958). From flow sheet to cost estimate. *Petroleum Refiner* 37:331–334.

Hand, W. E. (1992). Estimating capital costs from process flow sheets. *Cost Engineers' Notebook*. Morgantown, WV: AACE International.

Humphreys, K. K., Wellman, P. (1996). *Basic Cost Engineering*. 3rd ed. New York: Marcel-Dekker, Inc.

Lang, H. J. (1948). Simplified approach to preliminary cost estimates. *Chem. Eng.*, 55:112–113.

Wroth, W. F. (1960). Factors in cost estimation. *Chem. Eng.*, 67.

RECOMMENDED READING

Aspen Richardson's General Construction Estimating Standards. Cambridge, MA: Aspen Technology, Inc. (Published annually on CD-ROM).

Aspen Richardson's Process Plant Construction Estimating Standards. Cambridge, MA: Aspen Technology, Inc. (Published annually on CD-ROM).

Bent, J. A. (1996). Humphreys, K. K., ed. *Effective Project Management Through Applied Cost and Schedule Control*. New York: Marcel Dekker.

Clark, F. C., Lorenzoni, A. B. (1997). *Applied Cost Engineering*. 3rd ed. New York: Marcel Dekker.

Cost Engineers' Notebook. (Various dates). Morgantown, WV: AACE International.

Humphreys, K. K., ed. (1991). *Jelen's Cost and Optimization Engineering*. 3rd ed. New York: McGraw-Hill Inc.

Humphreys, K. K., Wellman, P. (1996). *Basic Cost Engineering*. 3rd ed. New York: Marcel Dekker.

Malstrom, E. (1981). *What Every Engineer Should Know About Manufacturing Cost Estimating*. New York: Marcel Dekker.

MEANS Building Construction Cost Data. (Published annually). Kingston, MA: R.S. Means Co.

MEANS Electrical Cost Data. (Published annually). Kingston, MA: R.S. Means Co.

MEANS Labor Rates for the Construction Industry. (Published annually). Kingston, MA: R.S. Means Co.

MEANS Mechanical and Electrical Cost Data. (Published annually). Kingston, MA: R.S. Means Co.

MEANS Repair and Remodeling Cost Data. (Published annually). Kingston, MA: R.S. Means Co. (various dates).

MEANS Site Work Cost Data. (Published annually). Kingston, MA: R.S. Means Co.

MEANS Square Foot Costs. (Published annually). Kingston, MA: R.S. Means Co.

MEANS Systems Costs. (Published annually). Kingston, MA: R.S. Means Co.

Note: All of the above listed R.S. Means Co. books and several other books from the R. S. Means Co. are available on one CD-ROM with built-in estimating capability under the title *Means CostWorks*.

4

Economic Equivalence and Profitability

4.1 INTRODUCTION

Through the following discussion of economic equivalence and profitability, the various terms used are defined in accordance with standard practice of the AACE International (see Appendix B). In this chapter all expended sums of money are indicated by a negative sign, and income is indicated by a positive sign.

Project evaluation requires that many angles be explored:

- Are raw materials available in sufficient quantities?
- Are the raw materials available from more than one source?
- Has the price of the raw materials risen a great amount over a short period of time?
- Is the cash requirement for the project available from current funds or is the cash required available as a loan at a reasonable rate of interest?
- Are the uses for the product known or will research to establish uses be one of the requirements for sale of the product in quantities to be produced?
- Is the product now manufactured by any other corporation?
- Will plants or machinery be costly to maintain?
- Are standard machines for producing the product available?
- Are exotic materials of construction required in piping, tank, or conveyor construction?
- Are patent rights available?
- Is the type of labor available that is required for production?

Most of the information for this chapter was obtained from a booklet "Discounted Cash Flow for Investment Decision," published by the Financial Publishing Company, supplemented with an article by John Rodgers for Reliance Universal Company, Houston, Texas, and also supplemented with examples and case studies by Julian A. Piekarski.

- Is the manufacturing process compatible with the present processes, and does the present management have enough experience to direct the operations?
- If a large amount of construction is necessary, can the contractors complete the project on schedule?

As was explained in Chapter 1, *simple interest* is a charge applied only to the amount of the debt for a single time period. As an example, suppose that $100 was borrowed for 2 years at 6% per year. Then at the end of 2 years $112 would be due. However *compound interest* is interest credited to the principal each period, and interest is earned on interest. The future value of a present amount A for N periods at an interest rate of I per period (expressed as a decimal) is

$$S = A(1 + i)^n$$

In engineering economics, since a year is usually taken as the period of time, compounding is done once a year. Increasingly some analysts are using continuous interest (see discussion of continuous interest in Chapter 1).

Sinking fund accounts can be set up into which a fixed amount is paid at the end of each period that, with the accumulated interest, amounts to a prescribed sum at a set time. An advantage of a sinking fund is that cash can be provided over a longer period of time to eliminate the possible difficulty of providing the entire amount at once. One disadvantage of a sinking fund is that the rate of return obtained in the sinking fund may be less than the rate of return that might be realized by reinvestment within the company.

Let R be a uniform end-of-year payment to a sinking fund and let S be the amount required at the end of n years at i per year expressed as a decimal. Then by item 5 in Table 1.1:

$$S = R\frac{(1 + i)^n - 1}{i} = RF_{RS,i,n} = RF_{RP,i,n}F_{PS,i,n}$$

Example 4.1

Replace a forklift truck costing $25,000 at the end of 5 years with a sinking fund yielding 5% compounded annually.

Solution   To find R, the end-of-year payment to the sinking fund, the relationship above gives

$$\$25,000 = R\frac{(1 : 05)^5 - 1}{0.05} = \$4524$$

That is, $4524 at the end of each year will yield $25,000 when compounded annually at 5% for 5 years.

If the firm had been netting 15% per year on the investment, then

$$S = R\frac{(1+i)^n - 1}{i} = \$4524\frac{(1.15)^5 - 1}{0.15} = \$30,503$$

That is, $30,503 would have been realized instead of $25,000. Thus there would be gain of $30,503 − $25,000 = $5503 by reinvestment in the firm.

Most firms can borrow cash using a delay payment period at the front of the payment schedule; only the interest is paid for the first few years, and after that the payments are started on the principal. In terms of the cash flow position, it may be advantageous in some instances to delay payment on the principal until cash can be realized on the investment; sometimes enough can be realized to have a profit above the interest paid.

4.2 PROFITABILITY ANALYSIS

There are several ways to evaluate the profitability of a project:

1. Percent return on original investment (ROI)
2. Payout time, the time required to recover the investment
3. Discounted cash flow rate of return (DCFRR), also known as interest rate of return, profitability index, investor's method, or true rate of return
4. Net present value (NPV) or "venture worth," which uses a fixed rate of return but reduces value at the prescribed rate of return.

All four of these techniques are discussed and illustrated in Chapter 1.

The payout time method has two weaknesses. First, only the payout period is considered when calculating the earnings; second, the payout method fails to consider the pattern of earnings.

Example 4.2

	Project		
Year	A	B	C
0 Investment	$− 100,000	$− 100,000	$− 100,000
1 Cash return	50,000	50,000	10,000
2 Cash return	50,000	50,000	90,000
3 Cash return	50,000	50,000	50,000
4 Cash return	50,000	50,000	30,000
5 Cash return	0	50,000	0

The payout time for each of projects A, B, and C is 2 years, making the projects equally attractive. However, when we analyze the data we find that project

A fails to show a profit after year 4. Project C recovers only $10,000 the first year, so $40,000 of profit or cash flow is delayed until the second year; thus $40,000 cannot be utilized in project C. Only project B continues to show a profit after year 4.

Another disadvantage of projects A and C is the amount of total earning. Project A earns $200,000 total, project C earns $180,000, and project B earns $250,000 for the same period and continues to earn.

A DCFRR calculation for the three projects yields

Project A	35%
Project B	41%
Project C	26%

Project B is the most attractive by that criterion.

Many factors enter into and determine the rate of return for projects:

- Risk of sale of the product and the price of the product
- Increase in cost of raw material and its availability
- Competition against superior or lower priced products
- Maintenance expenses, if excessive

Sales volume and selling price are particularly important in profitability studies since they can make or break a project. The ratio of profit to sales is an important figure in finance. Thus if sales are $184,300,000 and profit is $18,280,000, the ratio is

$$\frac{18.28}{184.3} = 9.91\%$$

That is, it takes almost $10 of sales to generate $1 of profit.

4.2.1 Depreciation

Depreciation methods were discussed in Chapter 1. The methods covered were:

- Straight line (SL)
- Declining balance (DB)
- Sum-of-the-years digits (SD)
- Double declining balance (DDB)
- Units of projection (UP)
- Accelerated cost recovery system (ACRS)
- Modified accelerated cost recovery system (MACRS)

The methods give different depreciation schedules, as the following example will show.

Example 4.3

Suppose that a machine costs $100,000 when new and lasts 16 years with no salvage value. For straight line depreciation the depreciation will be constant each year at

$$\frac{\$100,000}{16} = \$6250 \text{ per year}$$

For the sum-of-the-years-digits method of depreciation, noting that the sum of the digits from $1 + 2 + 3 + \cdots + 16$ is 136, the results are

Year	Depreciation		Percent
1	$100,000 \times (16/136) =$	$11,800	11.8
2	$100,000 \times (15/136) =$	11,000	11.0
3	$100,000 \times (14/136) =$	10,300	10.3
4	$100,000 \times (13/136) =$	9,600	9.6
⋮	⋮		⋮
14	$100,000 \times (3/136) =$	2,200	2.2
15	$100,000 \times (2/136) =$	1,500	1.5
16	$100,000 \times (1/136) =$	700	0.7
	Cumulative $= \$100,000$		

For the double declining balance method, the factor is:

$$\frac{2}{16} = 0.125$$

and the results are:

Year	Book value	Depreciation	Percent
1	$100,000	$100,000 \times 0.125 = \$12,500$	12.5
2	87,500	$87,500 \times 0.125 = 10,938$	10.9
3	76,562	$76,562 \times 0.125 = 9,570$	9.6
4	66,992	$66,992 \times 0.125 = 8,374$	8.4
⋮	⋮	⋮	⋮
14	15,422	$15,422 \times 0.125 = 1,928$	2.2
15	13,494	$13,494 \times 0.125 = 1,687$	1.9
16	11,807	$11,807 \times 0.125 = 1,476$	1.7
		Cumulative $= \$89,669$	

The double declining balance method gives an accumulated depreciation of $88,193, leaving $10,331 not depreciated. This circumstance can be avoided by shifting to straight line depreciation at any chosen point. Thus suppose that the shift to straight line is made at the end of year 9 when the book value, the unde-preciated value, is $34,362. Then the straight line depreciation with 8 years remaining is $34,362 ÷ 8 = $4295, and the tabulation becomes:

Year	Book value	Double declining balance depreciation	Straight line depreciation	Cumulative depreciation
1	$100,000	$12,500		$12,500
2	87,500	10,938		23,538
3	76,562	9,570		33,008
⋮	⋮	⋮		⋮
8	39,270	4,908		65,638
9	34,362	└────────→	$4,295	69,933
10	30,067		4,295	74,228
⋮	⋮	⋮	⋮	⋮
14	12,887		4,295	91,408
15	8,592		4,295	95,703
16	4,297		4,297	100,000

Thus a switch to the straight line method of depreciation permits the entire invest-ment of $100,000 to be written off.

The previous calculations neglected salvage value. For U.S. capital invest-ments made prior to 1981, when salvage value was included in depreciation cal-culations, it was never permissible to depreciate more than the depreciable value, which is first cost less salvage value. If the project was sold for its salvage value at the end of the project, there was no gain or loss at that time and no tax consider-ation. If it was sold for more than the salvage value, there was a profit subject to tax. Conversely, if it was sold for less than the salvage value, there was a loss that reduced the tax.

For United States capital investments made in 1981 and subsequent years, ACRS or MACRS, which do not consider salvage at all, apply. These methods are described in Chapter 1 as is the units of production method.

4.2.2 Practical Examples of Economic Comparisons and Profitability

Example 4.4   Adding to Existing Facility

Existing facility 1 generates a cash flow of $80,000 per year and an addition to this facility at a cost of $100,000 is contemplated. The existing facility realizes 15% on investment, and the addition will generate an increase of $21,000 cash flow for a period of 16 years. The net present value, or present worth of the addition and the extra income, is (by Chapter 1, Table 1.1)

$$\text{NPV} = -100,000 + 21,000 F_{RP,i,16}$$

$$= -100,000 + 21,000 \frac{(1+i)^{16} - 1}{i}$$

For the net present value to be zero, i becomes the discounted cash flow rate of return and

$$0 = -100{,}000 + 21{,}000 F_{RP,i,16}$$

$$F_{RP,i,16} = 4.76 = \frac{(1+i)^{16} - 1}{i}$$

The value of i can be solved by trial and error and is about 20%. Also one can scan tables for $F_{RP,i,16}$ and find that at 20% $F_{RP,20\%,16} = 4.73$, which is close enough. Thus the addition earns 20% rate of return or 5% more than that for the existing facility.

If the income per year is irregular, as opposed to uniform, the present worths for each year must be calculated individually and summed using a trial value for i until the present worth is zero. The better pocket programmable calculators and many personal computer software packages have programs for finding the DCFRR so that the calculation no longer has to be done by hand.

Example 4.5   Maintenance Decision

Discounted cash flow can be applied in a similar manner to evaluate maintenance decisions. Consider replacing a pump when two different types can be purchased and either type will perform the task.

Pump 1   Costs $13,000 with a life of 18 years with negligible maintenance.

Pump 2   Costs $10,000 with a life of 18 years and a total maintenance of $7000 (70% of original cost) split between the ends of years 6 and 12.

Let the tax rate be 34%. The pumps are used for replacement; assume additionally that the first cost can be written off immediately for tax purposes. Similarly the maintenance cost of $7000 per year can be written off immediately for tax purposes.

Solution   The present values of pump 1 over pump 2 can be shown in a tabulation: Note that the after-tax *extra* cost for pump 1 is

$$\$13{,}000(1 - 0.34) - \$10{,}000(1 - 0.34) = \$1980$$

The maintenance cost after taxes is

$$\$3500(1 - 0.34) = \$2310$$

The discount factor is $(1 + i)^{-n}$ and can be obtained from tables or calculated as needed. The tabulation is:

		Discount factor		Total discounted cash flow	
Year	Extra cash flow	10%	12%	10%	12%
0	$-1980	1.0000	1.0000	$-1980	$-1980
6	+2310	0.5645	0.5066	+1304	+1170
12	+2310	0.3186	0.2567	+736	+593
				$+60	$-217

By interpolation the discounted cash flow is zero at about 10.4%. That is, pump 1 and pump 2 are equivalent choices at a return rate of 10.4%. Therefore, if the time value of money (the hurdle rate) to the company is less than 10.4%, pump 1 is the best choice. Otherwise, pump 2 should be chosen.

Example 4.6   Population Growth

A person starts a rabbit farm with 1000 rabbits. Rabbits increase their population by 10% each year. Each year 50 rabbits must be sold to pay expenses. How long will it take to reach a population of 4000 rabbits?

Solution   Set 500 rabbits aside to generate the 50 rabbits per year sold for expenses: The other 500 rabbits must increase sevenfold to 3500 at 10% per year for n years. From tables or by trial and error, $(1.10)^n = 7$ and $n = 20.4$ years.

Example 4.7   Immigration

Assume that 300,000 new people enter a country each year, and after entering their population increases by 1% per year. If none die, how many people will be added from immigration in a period of 50 years?

Solution   Using the methods of Chapter 1, Table 1:1,

$$S = RF_{RS,1\%,50} = R\frac{(1 + i)^n - 1}{i} = 300,000\frac{(1.01)^{50} - 1}{0.01}$$

$$= 19,338,900 \text{ people at end of 50 years}$$

Example 4.8   Inventory Turnover

A store has a net profit of 5% on each sale and an average turnover rate of 3 times a year. If the turnover increases to 5 times a year, how will the profit be increased?

Solution

At 3 turnovers/year	$(1.05)^3 = 1.1576$	15.76% profit
At 5 turnovers/year	$(1.05)^5 = 1.2763$	27.63% profit

The profit would increase from 15.76% to 27.63%.

Example 4.9   Rate of Growth

A corporation estimates a market for 5,000,000 units per year of a product. The present manufacturing capacity of the plant is 150,000 units per year, and the corporation wants to expand it to 5,000,000 units per year within 17 years (the life of the process patent), what will be the rate of expansion per year?

Solution   If 150,000 must grow to 5,000,000, then 1 must grow to $5,000,000/150,000 = 33:33$. We can scan tables and find that at 22%, 1 grows to 29.38 in 17 years, and at 23%, 1 grows to 33.75 in 17 years. Thus the rate of growth is about 23% per year.

Example 4.10   Pollution

The present level of an impurity in the water downstream of a plant is 100 units per gallon, and the allowable level is 300 units per gallon. Each piece of added equipment will increase the amount of pollution 2%. How many pieces of additional equipment can be added?

Solution

$$(1.02)^n = \frac{300}{100} \quad \text{or} \quad n \log 1.2 = \log 3$$

for which $n = 55.5$. Thus 55 new pieces of equipment can be added without exceeding the allowable pollution level.

4.2.3   Lease Versus Purchase

The question of when to lease and when to purchase equipment becomes very complex at times, and all aspects must be considered. Following are some facts to note before *purchasing* equipment.

1.   The effect of the added capital assets on the tax structure of the corporation must be determined, as well as the additional bookkeeping necessary to calculate depreciation, and so on, when the assets are acquired.

2.   Records of maintenance and efficiency of the different types of equipment must be maintained so that the time of replacement can be determined and provided for in the budget.

3.   Storage and repair facilities must be provided for the equipment and the repair parts. The availability of trained maintenance personnel must be evaluated and the time they will be required must be determined.

4. One of the most important considerations must be to determine how frequently the equipment will be employed in the normal working year. This would include evaluating the mobility of the equipment and the ability to utilize the equipment in other locations of the mill or at other plant sites, and how the charges for the equipment would be distributed.

5. Security requirements in the areas in which the equipment or repair parts and tools are stored can be an important consideration.

6. If the equipment is available on the site when required, a move-in/ move-out charge is eliminated.

Before *leasing* equipment, one should consider the following:

1. No capital assets are added to affect the tax structure of the corporation, and bookkeeping is not increased to calculate depreciation, and so on; maintenance of records is minimized.

2. No formal records are needed for normal maintenance and efficiency determination.

3. Storage and repair facilities need not be provided; trained maintenance personnel may not be needed.

4. If a type of equipment is not employed frequently or for long periods of time, the equipment rental will cease when the equipment is released from the project. Also the charge for the equipment can be transferred by the contractor to another project with ease. If the equipment is owned, internal bookkeeping and agreement on the amount of charges are required.

5. Security of the equipment must be maintained only when the equipment is utilized since the equipment is on site only when required.

6. Availability of the equipment and the distance of the haul must be considered. A contractor who can furnish the required equipment with reasonable move-in/move-out charges must be selected.

Example 4.11   Lease vs. Purchase

A new welding machine (gasoline/electric) has been leased for $300 per month, and a new machine can be purchased for $2500. We must decide whether to purchase a new machine or to continue to lease a machine. The data are as follows:

Cost of new machine, $	2500
Cost of leasing, $/yr	3600
Expected life of machine/yr	8
Gas, oil, maintenance, etc., is canceled since they must be provided for both machines	0
Straight line depreciation, total years	8
Salvage value	0
Tax rate, %	34

Solution   The leasing cost can be written off at once for tax purposes and after taxes is

$3600(1 − 0 : 34) = $2376

The new machine generates a depreciation expense of $2500/8 = $312.50 which, after taxes, becomes

$312.50(1 − 0.34) = $206.25

The extra cash flow by purchasing is

$−2500 at zero time

and there is a savings over leasing at the end of each of the 8 years amounting to

$2376 + $206.25 = $2582.25

A year-by-year tabulation is not necessary in this particular example but is used here as a guide to the general procedure. The discounted cash flow for the extra cash flow generated by purchase of the machine is:

Year	Extra cash flow $	Discount factor 90%	Discount factor 80%	Total discounted cash flow $ 90%	Total discounted cash flow $ 80%
0	−2500	1.0000	1.0000	−2500	−2500
1	+2582.25	0.5263	0.5556	1359	1345
2	+2582.25	0.2770	0.3086	715	797
3	+2582.25	0.1458	0.1715	376	443
4	+2582.25	0.0767	0.0953	198	246
5	+2582.25	0.0404	0.0529	104	137
6	+2582.25	0.0213	0.0294	55	76
7	+2582.25	0.0112	0.0163	29	42
8	+2582.25	0.0059	0.0091	15	23
				$−351	$+699

By interpolation of the discount rates to a net cash flow of $0, the discounted cash flow rate of return is about 86.7% in favor of purchasing a new machine.

4.2.4   Economics for Capital Projects or Replacements

Economic justification for construction or replacement of a project should be viewed from the impact of the cost of the project on the method of obtaining revenue as well as from other aspects. Since most corporations obtain revenue from the sales of products, the impact of the cost on sales revenue must be considered.

When calculating discounted cash flows, the cost of money, profit, depreciation, and so on, must be established to a high degree of accuracy. These figures can then be applied in other methods of calculating risk and the profitability of projects.

One of the first steps in determining profitability in the impact on sales method is to examine the present percentage of net profit to sales. This can be done by dividing the net profit for the year by the sales volume in dollars for the year. If we assume the sales volume to be $184,300,000 and the net profit to be $18,280,000, we find the percentage to be

$$\frac{\$18,280,000}{\$184,300,000} = 9.91\%$$

By applying this percentage, the amount of sales required to offset the expenditure can be found. For each dollar expended, the amount of sales required is found to be

$$\frac{\$1}{0.0991} = \$10.09$$

This calculation illustrates the importance of the corporation saving comparatively small amounts of money as well as saving larger amounts. When viewed from the evaluation aspect and applied to the project, several determinations must be made:

1. The amount of sales volume in dollars per year that will be generated
2. The amount of net profit realized from the additional sales volume
3. The amount of operating expense to be eliminated by more efficient operations
4. The duration in time of the increase in production or duration in time of the savings in expense
5. The amount of sales volume in dollars per year required to offset the expenditure

Example 4.12

Consider the following project:

Estimated ratio net profit to sales	= 10%
Estimated cost of project	= $45,000,000
Estimated sales volume required	$= \dfrac{\$45,000,000}{0.10}$
	= $450,000,000 total
Estimated sales volume generated/year	= $37,000,000
Estimated net profit generated/year	= $3,700,000

$$\text{Depreciation (16 years, straight line)} = \frac{\$45,000,000}{16}$$

$$= \$2,812,500$$

$$\text{Breakeven point (net profit sales only)} = \frac{\$45,000,000}{\$3,700,000}$$

$$= 12.16 \text{ years}$$

$$\text{Breakeven point (net profit} + \text{depreciation)} = \frac{\$45,000,000}{\$3,700,000} + \$2,812,500$$

$$= 6.9 \text{ years}$$

When the discounted cash flow method is applied using the same 10% factor to determine if the project is feasible, we find:

Year	Cash flow	Discount factor interest rate: 10%	Total discounted cash flow: 10%
0	($-45,000,000)	1.0000	($-45,000,000)
1	6,512,500*	0.9091	$5,920,514
2	6,512,500	0.8264	5,381,930
3	6,512,500	0.7513	4,892,841
4	6,512,500	0.6830	4,448,038
5	6,512,500	0.6209	4,043,611
6	6,512,500	0.5645	3,676,306
7	6,512,500	0.5132	3,342,215
8	6,512,500	0.4665	3,038,081
9	6,512,500	0.4241	2,761,951
10	6,512,500	0.3855	2,510,569
11	6,512,500	0.3505	2,282,631
12	6,512,500	0.3186	2,074,883
13	6,512,500	0.2897	1,886,671
			$46,260,241
14	6,512,500	0.2633	1,714,741
15	6,512,500	0.2394	1,559,093
16	6,512,500	0.2176	1,417,120
			$50,951,195

*Net profit + depreciation = $3,700,000 + $2,812,500 = $6,512,500.

As shown above, the project would earn the required 10% profit and recover the capital investment in 13 years, with $1,260,241 or 2.8% of the initial $45,000,000 investment as a surplus. If the calculation is completed through the sixteenth year, when the project has been totally depreciated, the

10% profit would be earned and the initial investment recaptured with a $5,951,195 surplus.

The example above illustrates a good investment. However, if money were expended but contributed nothing to the profit or did not reduce operating expenses, we would have a different situation.

Example 4.13

Money expended	=$150,000
Sales needed to break even	=$1,500,000
Estimated ratio net profit to sales	=10%

When this example is viewed in terms of the effect on sales, profit, and the loss of revenue that could have been realized if the profit had been invested in a project yielding 10% profit, an even greater impact on earnings is noted.

$$\$150,000 + (0.10 \times 150,000) = \$150,000 + \$15,000 = \$165,000$$

$$\frac{\$165,000}{0.10} = \$1,650,000 \text{ in additional sales required to break even}$$

When viewed using the calculation above, we find the sales required to offset the additional expenditure are $1,650,000. When this is applied to a corporation earning 10% on sales of $182,800,000 or $18,280,000, the impact of the expenditure becomes:

$$\frac{\$18,280,000}{\$182,800,000} = 10\% \text{ net profit}$$

$$\frac{\$165,000}{\$18,280,000} = 0.009$$

where 0.009 equals approximately 1% of the total net profit of the corporation.

The ratio of the smaller gain in sales can present another picture, as can be seen above: $1,650,000 of the sales revenue must be expended for the $150,000 spent. This represents

$$\frac{\$1,650,000}{\$182,800,000} = 0.009$$

where 0.009 equals approximately 1% of total sales.

If a corporation with a sales volume of $182,800,000 has a healthy net gain in sales of 12% for the year, or $21,936,000, an expenditure of only $150,000 for a nonproductive project would result in a reduction of

$$\frac{\$1,650,000}{\$21,936,000} = 0.075$$

where 0.075 equals 7.5% of the net gain in sales.

This would for all practical purposes reduce the net gain in sales to 11% ($20,108,000) from the 12% ($21,936,000) gain, as 7.5% of the additional sales would be offset by the nonproductive expenditure.

The examples above illustrate the importance of the evaluation of all projects to be undertaken by corporations. The examples also illustrate the importance of efficient operations and budget planning when the impact on sales ratio of profits and the impact on net profits from a comparatively small unnecessary expenditure are noted.

When considering the growth of a corporation, a 12% increase in sales appears to be a healthy rate of growth. However, when $150,000 of unnecessary expenditures can lower 1% of the sales growth and $21,936,000 \times 10\% = \$2,193,600$ can negate the entire growth in sales, the careful evaluation of expenditures can be seen as being of great importance.

The importance of implementing more efficient conditions or methods wherever possible can be seen readily by the fact that a saving of $150,000 in operating expense is the equivalent of a 1% growth in sales for a corporation with a sales volume of $182,800,000. This becomes even more significant if the operating expenses are average, or approximately 85% of sales or revenue.

$$0.85\% \times \$182,800,000 = \$155,380,000$$

When the realization that, with a comparatively small savings,

$$\frac{\$150,000}{\$155,380,000} = 0.009$$

(0.009 equals approximately 1/10 of 1% of operating expense), the equivalent of a 1% growth in sales is realized, and the savings become even more important.

4.2.5   Scheduling a Loan Payment

A loan of $1,000,000 is borrowed with interest at 8% per year or 2% per quarter. The loan will be repaid over a scheduled period of 25 years. During the first 5 years, or 20 periods, only the interest will be paid. During the next 20 years, or 80 periods, the loan balance will be reduced uniformly and interest paid on the loan balance.

Here $1,000,000 is repaid uniformly over the last 80 periods at $1,000,000/80 = $12,500. The payment schedule is:

Quarter	Interest per period	Principal per period	Payment per period	Loan balance
1–20	$20,000	$0	$20,000	$1,000,000
21	20,000	12,500	32,500	987,500
22	19,750	12,500	32,250	975,000
⋮	⋮	⋮	⋮	⋮
50	12,750	12,500	25,250	625,000
51	12,500	12,500	25,000	612,500
⋮	⋮	⋮	⋮	⋮
99	500	12,500	13,000	12,500
100	250	12,500	12,750	0
	$1,210,000	$1,000,000	$2,210,000	

At times it might be advantageous to delay payments toward principal to maintain a better cash flow position, and the amount paid in interest for the delay period must be weighed in this decision. In this example, the interest for the 5 years is 49.38% of the interest of the remaining 20 years of the loan. The interest for the first 5 years is also 40% of the loan. If the loan was $50,000,000, the interest for the 5-year delay period would be $20,000,000, or quarterly payments of $1,000,000 totaling $4,000,000 per year.

The $20,000,000 paid over the 5-year period would then be an addition to the $50,000,000 cost of the project and for discounted cash flow calculations, the $20,000,000 would be paid by the profit from the project for the first 5 years. Any cash expended while the project was under construction would be included with the loan.

Example 4.14 Loan Schedule

For the installation of a paper machine and stock preparation area, the following data apply:

Total installed cost	$55,000,000
Amount of loan	$50,000,000
Cash outlay during construction	$5,000,000
Repayment of loan: first 5 years' interest only; thereafter paid off in 80 quarterly payments in beginning in the sixth year	

Interest rate, %	8
Depreciation method	Straight-line
Depreciation years	16
Depreciation, $/yr: 55,000,000/16	$3,437,500
Net profit after interest and loan payments	
Year 1, $/yr	$1,500,000
Years 2–25, $/yr	$3,300,000
Cash flow = depreciation + net profit	
Year 1, $3,437,500 + $1,500,000	$4,937,500
Years 2–16, $3,437,500 + $3,300,000	$6,737,500
Years 17–25, 0 + $3,300,000	$3,300,000

The first part of the cash flow calculation is a restricted view to accentuate paying only interest on the loan for the first 5 years. The project is judged from a commitment of $20,000,000 in interest (5 years at $4,000,000/yr). Although it is true that the interest will not be due at zero time but will be spread out in the future, the view here is to consider the $20,000,000 as a present commitment of the project. For the first 5 years the calculation is:

Year	Reference	Cash flow	Discounted at 17%	Discounted at 18%
0	$-20,000,000		$-20,000,000	$-20,000,000
1		$4,937,500	4,220,081	4,184,531
2		6,737,500	4,921,744	4,838,873
3		6,737,500	4,206,895	4,100,443
4		6,737,500	3,595,804	3,475,203
5		6,737,500	3,072,974	2,944,961
			$+17,498	$-455,989

The present worth will be zero at about 17.04%. This says in effect that if the total interest payments for the first 5 years, i.e., $20,000,000, were swapped for the first 5 years of the project, the discounted cash flow rate of return would be about 17.04%.

The second part of the cash flow calculation goes from years 6 through 25, and the reference point here is a $55,000,000 investment. All discounting is done with respect to the beginning of year 6. The calculation is:

Year	Reference	Cash flow	Discounted at 8%	Discounted at 10%
	$-55,000,000		$-55,000,000	$-55,000,000
6		$6,737,500	6,238,251	6,125,061
7		6,737,500	5,776,059	5,567,870

Year	Reference	Cash flow	Discounted at 8%	Discounted at 10%
⋮		⋮	⋮	⋮
15		6,737,500	3,120,810	2,597,306
16		6,737,500	2,889,714	2,361,494
17		3,300,000	1,310,430	1,051,380
18		3,300,000	1,213,410	956,010
⋮		⋮	⋮	⋮
24		3,300,000	764,610	539,550
25		3,300,000	707,850	490,380
			$+1,939,281	$-4,579,217

The present worth will be zero at about 8.60%. This says that if the project is regarded as an investment of $55 million at the beginning of the sixth year, it will have a discounted cash flow rate of return of 8.60% from the period starting at the beginning of the sixth year and ending at the end of the twenty-fifth year.

One of the major advantages in the process of delaying the payments toward the loan principal is to take advantage of the cash flow realized from depreciating the capital assets during the time. There is a possibility, however, that the delay of payments will be a detriment if the debt position of the corporation is good (very few debts outstanding), and a major expansion or major maintenance program is contemplated within a 4- to 6-year period. A possibility exists under these circumstances that the credit rating of the corporation would be affected. The corporation may then be in a better position to repay the debt at a more accelerated rate to ensure a more advantageous position in credit rating and solvency. This may ensure the ability to borrow sufficient funds at a later date to allow expansions or maintenance at a critical time in the future.

When contemplating the life of a corporation, plans must be projected to allow the corporation to continue indefinitely and not be affected by changes in personnel or any other short-term occurrence. Cycles in the change of personnel or business climate (cost of money expenses, etc.) should be projected and planned so that the corporation will have a continual healthy growth.

Example 4.15   Profitability of a Paper Machine

A paper machine has a production rate of 375 tons/day for 350 days/yr with paper selling at a price of $281.50 per ton. Gross income per year is

$$375 \times 350 \times \$281.50 = \$36,946,875 \text{ per year}$$

After expenses, but before taxes, only 15.16% remains or

$$0.1516 \times \$36,946,875 = \$5,601,146 \text{ per year}$$

After allowing $2,238,980 for taxes, the net income after taxes is

$5,601,146 − $2,238,980 = $3,362,166 per year

The capital cost for the machine is $34,000,000. A 10% tax credit is available, and the machine will be depreciated at straight line depreciation over 16 years or at

$$\frac{\$34,000,000}{16} = \$2,125,000 \text{ per year}$$

The cash flow is the sum of net profit after taxes plus depreciation and is

$3,362,166 + $2,125,000 = $5,487,166 per year

The return on original investment for the machine is

$$\frac{\$5,487,166}{\$34,000,000} = 16.1\%$$

The payback time will be

$$\frac{\$34,000,000}{\$5,487,166} = 6.20 \text{ years}$$

All the calculations above have neglected a 10% tax credit which amounts to

$34,000,000 × 0.10 = $3,400,000

When the 10% tax credit is considered, the return on original investment becomes

$$\frac{\$5,487,166}{\$34,000,000 - \$3,400,000} = 17.9\%$$

The payback time when considering the tax credit is

$$\frac{\$34,000,000 - \$3,400,000}{\$5,487,166} = 5.58 \text{ years}$$

Cash flow for the first year is

Profit after tax	$3,362,166
Depreciation	2,125,000
10% investment tax credit	3,400,000
	$8,887,166

and for the other years is

Profit after tax	$3,362,166
Depreciation	2,125,000
	$5,487,166

Calculating venture worth

Year	Amount	10% present worth factor	Present worth
0	$-34,000,000	1.0000	$-34,000,000
1	8,887,166	0.9090	8,078,434
2	5,487,166	0.8264	4,534,594
3	5,487,166	0.7513	4,122,508
⋮	⋮	⋮	⋮
15	5,487,166	0.2394	1,313,628
16	5,487,166	0.2176	1,194,007
			$+12,035,904

Thus venture worth of the project at 10% rate of return is $+12,036,000.

The next problem departs from the usual assumptions of uniform costs and replaces it with realistic values that change each year. All this change means is that the calculation of present worth and rate of return is a bit more complex, but it does not change the approach.

Example 4.16   Energy Conservation

Three years ago a pipe manufacturing company built and installed a gas drying oven at a total cost of $20,000. Operating costs of the oven have been less than satisfactory, and a task force has been appointed to find ways to reduce the excessive fuel costs. A firm manufacturing drying ovens of a new and radically different design has approached the company with a model that sells for $40,000 and promises to reduce costs, particularly energy costs, very dramatically. The task force has been asked to determine whether this new oven can be justified by the potential savings in fuel and maintenance costs. Details of the alternatives are tabulated below:

	Existing	New
Investment	$20,000	$40,000
Life, years	10	10
Age, years	3	—
Depreciation	MACRS	MACRS
Book value	$5,760	—
Salvage	0	0
Income taxes (state and federal)	40%	40%
Cost of capital	15%	15%

	Comparative Costs			
	Existing		New	
Year	Fuel	Maintenance	Fuel	Maintenance
1	$12,000	$600	$6000	—
2	12,100	1000	6000	400
3	12,200	1200	6100	600
4	12,300	1400	6100	800
5	12,400	1800	6500	1000
6	12,500	2000	6500	1200
7	12,600	2400	6800	1400
8	12,700	3000	6800	1400
9	12,800	3500	6900	1500
10	12,900	4000	7000	1500

Assume that both ovens will last for 10 years and that the book value of the existing oven will be credited to the new oven, if acquired. The treatment of this write-off is identical to depreciation, i.e., since it is not a cash expense, it must be added back to the cash flow after taxes. Depreciation values are as follows:

	Existing			New	
Year Investment	MACRS %	Amount $20,000	Remaining book value	MACRS %	Amount $40,000
1	20.00	$4000	16,000	20.00	$8,000
2	32.00	6400	9,600	32.00	12,800
3	19.20	3840	5,760	19.20	7,680
4	11.52	2304	3,456	11.52	4,608
5	11.52	2304	1,152	11.52	4,608
6	5.76	1152	—	5.76	2,304
		$20,000			$40,000

Tables 4.1 and 4.2 show the respective cash flows for the new and existing furnaces, and Table 4.3 provides the incremental analysis.

Example 4.17   Revenue Requirements—Electric Utility

An electric utility is planning to add a 350 MW generating plant that will have a coal-fired steam generator designed to burn high sulfur coal shipped by unit train from a mine that is a wholly-owned subsidiary. The plant is to be designed for peaking service which means it will be at full load every weekday, down to

Table 4.1 New Furnace

	0	1	2	3	4	5	6	7	8	9	10
Fuel		6,000	6,000	6,100	6,100	6,500	6,500	6,800	6,800	6,900	7,000
Maintenance		—	400	600	800	1,000	1,200	1,400	1,400	1,500	1,500
Depreciation		8,000	12,800	7,680	4,608	4,608	2,304				
Retirement	5,760										
Total expenses	5,760	14,000	13,200	14,380	11,508	12,108	10,004	8,200	8,200	8,400	8,500
Taxable balance	(5,760)	(14,000)	(13,200)	(14,380)	(11,508)	(12,108)	(10,004)	(8,200)	(8,200)	(8,400)	(8,500)
Tax @ 40%	(2,304)	(5,600)	(5,280)	(5,752)	(4,603)	(4,843)	(4,002)	(3,280)	(3,280)	(3,360)	(3,400)
Balance, after tax	(3,456)	(8,400)	(7,920)	(8,628)	(6,905)	(7,265)	(6,002)	(4,920)	(4,920)	(5,040)	(5,100)
Plus:											
Depreciation		8,000	12,800	7,680	4,608	4,608	2,304				
Retirement	5,760										
Less:											
Investment	(40,000)	(400)									
Cash flow	(37,696)	(400)	4,880	(948)	(2,297)	(2,657)	(3,698)	(4,920)	(4,920)	(5,040)	(5,100)

Present worth @ 15% = −$45,362; total cash flow = (62,796).

Table 4.2　Existing Furnace

	0	1	2	3	4	5	6	7	8	9	10
Fuel		12,000	12,100	12,200	12,300	12,400	12,500	12,600	12,700	12,800	12,900
Maintenance		600	1,000	1,200	1,400	1,800	2,000	2,400	3,000	3,500	4,000
Depreciation		2,304	2,304	1,152							
Total expenses		14,904	15,404	14,552	13,700	14,200	14,500	15,000	15,700	16,300	16,900
Taxable balance		(14,904)	(15,404)	(14,552)	(13,700)	(14,200)	(14,500)	(15,000)	(15,700)	(16,300)	(16,900)
Tax @ 40%		(5,962)	(6,162)	(5,821)	(5,480)	(5,680)	(5,800)	(6,000)	(6,280)	(6,520)	(6,760)
Balance, after tax		(8,942)	(9,242)	(8,731)	(8,220)	(8,520)	(8,700)	(9,000)	(9,420)	(9,780)	(10,140)
Plus:											
Depreciation		2,304	2,304	1,152							
Cash flow		(6,638)	(6,938)	(7,579)	(8,220)	(8,520)	(8,700)	(9,000)	(9,420)	(9,780)	(10,140)

Present worth @ 15% = −$40,448; Total cash flow = (84,935).

Table 4.3 Incremental Approach

	Existing furnace	New furnace	New minus existing
0	—	(37,696)	(37,696)
1	(6,638)	(400)	6,238
2	(6,938)	4,880	11,818
3	(7,579)	(948)	6,631
4	(8,220)	(2,297)	5,923
5	(8,520)	(2,657)	5,863
6	(8,700)	(3,698)	5,002
7	(9,000)	(4,920)	4,080
8	(9,420)	(4,920)	4,500
9	(9,780)	(5,040)	4,740
10	(10,140)	(5,100)	5,040
	(84,935)	(62,796)	22,139

Present worth @ 15% = −$4,914; DCFRR = 10.97%.

20% at night, and offline on weekends. The following data provide the information for the financial calculations:

Capital	= $300,000,000
Plant capacity factor	= 50%
Fuel cost	= $25/ton
	12,000 BTU/lb
Plant heat rate	9,500 BTU/kWh
Annual operating/maintenance expense	= $3000/MW
Cost of capital	= 10%
Income tax (state and federal)	= 40%

Energy generated at 50% capacity factor:

$$kWh/yr = 350\,MW \times 50\% \times 8760\,hr/yr \times 1000\,kW/MW$$
$$= 1,533,000,000\,kWh/yr$$

Fuel consumption:

$$\frac{(1.533 \times 10^9\,kWh/yr)(9.5 \times 10^3\,BTU/kWh)}{(12,000\,BTU/lb)(2000\,lb/ton)} = 606,813\,ton/yr$$

Fuel cost:

$$= 606,813\,ton/yr \times \$25/ton$$
$$= \$15,170,313/yr$$

	Cash Flow ($000)	
Year	0	1–15
Fuel cost	–	$15,170
O&M expense	–	1,050
Depreciation		20,000
Total expense	–	36,220
Taxable balance	–	(36,220)
Tax @ 40%	–	(14,488)
Balance after tax	–	(21,732)
Plus: depreciation	–	20,000
Less: investment	(300,000)	
Cash flow	(300,000)	1,732

NPV @ 10% = $-300,000 + 1732 $(F_{RP,10\%,15})$
= $-300,000 + (1732 \times 7.606)$
= $-286,826$

Revenue
Requirements = $(-286,826/-0.6).13147$ = $62,848/yr
= $0.041/kWh
= 41 mills/kWh

Example 4.18   Risk Analysis—Capital Cost Estimate*

The basic procedure used in this example was originally introduced by Cooper and Davidson (1976) and is fully explained in a paper presented by Piekarski (1984). Given the generation plant investment in Example 4.17, a construction firm has been awarded the job of erecting the power plant. A definitive estimate of capital cost was made as shown below along with the probable range of variation to the estimate:

Capital Investment 350-MW Plant (Million $)		
	Most Likely (ML)	Variation Range
Structures	37	(+10%, −10%)
Boiler plant	117	(+15%, −10%)
Turbine plant	44	(+15%, −10%)
Electrical	16	(+10%, −10%)
Flue gas desulfurization	46	(+20%, −10%)
Indirects	40	(+30%, −10%)
	$300	

---

*The manual risk analysis approaches described in Example 4.18 and 4.19 give approximate estimates of the probability of achieving project objectives. Excellent software programs are now available which greatly simplify the estimation of overrun probability and of the amount of required contingency to reduce the overrun probability to an acceptable level. This software is preferred to the manual approaches and provide more reliable answers. Chapter 9 discusses computerized risk analysis and contingency determination using Monte Carlo techniques.

Management wishes to determine the contingency that needs to be added in order to insure that there is no more than a 30% probability that the project will overrun the bid.

Solution   The cost estimate on the project utilizes the following equations to calculate the contingency:
To calculate the mean of each line item:

$$\text{Mean} = (H + 2 \times ML + L)/4$$

where

H = highest anticipated value
ML = most likely value
L = lowest anticipated value

To estimate the standard deviation of each line item:

$$SD = (H - L)/2.65$$

To calculate the overall mean and standard deviation:

$$\text{Mean} = \text{sum of the individual line item means}$$

$$\text{Total SD} = \sqrt{\text{sum of } (SD)^2}$$

Table 4.4 summarizes the computations of risk parameters.
The results are best explained by charts. The risks incurred in Example 4.17 are profiled in Fig. 4.1. The diagonal line of the chart is a cumulative probability distribution of the range of values calculated in Table 4.4. The crux of the chart is how it serves as a means to respond to the requirement that there be no more than a 30% chance that the project would overrun. A dashed line is superimposed at

Table 4.4   Risk Parameters

	Most likely	High	Low	Mean	Std. Dev.
Structures ($\pm 10\%$)	37	40.70	33.3	37.00	2.79
Boiler plant ($+15\% -10\%$)	117	134.55	105.3	118.46	11.04
Turbine plant ($+15\% -10\%$)	44	50.60	39.6	44.55	4.15
Electrical ($\pm 10\%$)	16	17.60	14.4	16.00	1.21
FGDS ($+20\% -10\%$)	46	55.20	41.4	47.15	5.20
Indirects ($+30\% -10\%$)	40	52.00	36.0	42.00	6.03
	$300			305.16	14.55

Range = $305.16 \pm 14.55 (\pm 5\%)$ = $290.61 to 319.71; use = $290 to 320.

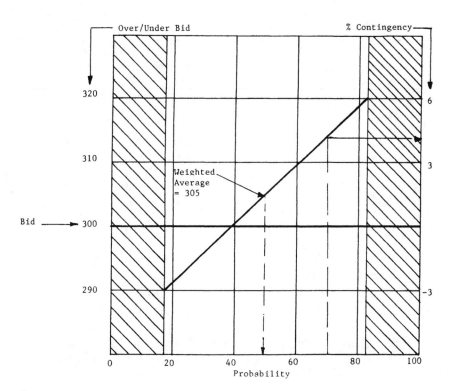

Figure 4.1 Contingency chart.

the 30% probability point which intersects the cumulative probability curve at 5% contingency (about $15M).

Example 4.19   Risk Analysis—New Product Evaluation

An electronic manufacturer is considering the introduction of a new product with the following financial characteristics:

Capital investment	$70,000
Depreciation	5 years, straight line
Sales potential	Approximately 7 times the capital investment over the life cycle ($500,000)
Variable cost	45 to 55% of sales
Income taxes (federal, state, and local)	40%
Cost of capital	12%, but the company has a minimum acceptable rate-of-return, the hurdle rate, of 20% for products in this risk category.

The cash flow table for the project is:

Year	0	1	2	3	4	5
		Cash Flow Table ($000)				
Revenues		50.0	71.5	90.0	127.5	160.0
Variable cost		27.5	39.2	49.5	58.7	73.6
Depreciation		14.0	14.0	14.0	14.0	14.0
Total cost		41.5	53.2	63.5	72.7	87.6
Taxable balance		8.5	18.3	26.5	54.8	72.4
Tax @ 40%		3.4	7.3	10.6	21.9	29.0
Balance, after tax		5.1	11.0	15.9	32.9	43.4
Plus: depreciation		14.0	14.0	14.0	14.0	14.0
Less: investment	70.0					
	−70.0	19.1	25.0	29.9	46.9	57.4
NPV @20% = 26.27	DCFRR = 33%					

As can be seen from the financial results, the project has a healthy net present value residual after 20% cost of money is deducted. Furthermore the rate of return of 33% is significantly above the hurdle rate of 20%.

Risk Analysis Using the Parameter Method   If we calculate the net present value of the revenues, variable costs, and fixed investment after taxes at 20% hurdle rate, the following values are obtained:

Net Present Value @ 20% (After Taxes) ($000)		
Revenues		$161.51
Variable cost	−$82.00	
Fixed investment	−$53.25	
Total costs	−$135.25	
Cash flow		$26.31

All this does is set the stage for the risk calculation using the parameter method. It can be seen that the net present value at this cash flow position corresponds to the previous calculation (except for minor differences due to rounding error). The following calculations utilize the formulas from Example 4.17.

Line item	Most likely	Low	High	Mean	Std. dev.	Range
		Calculation of Risk Parameters				
Revenue	161.51	129.2	177.7	157.5	18.3	−20% to +10%
Expense	82.00	73.8	102.5	85.1	10.8	−10% to +25%
Investment	53.25	47.9	71.9	56.6	9.1	−10% to +35%
Cash flow	26.31			15.8	23.0	
		Range: 16 ± 23 = −7 to +39				

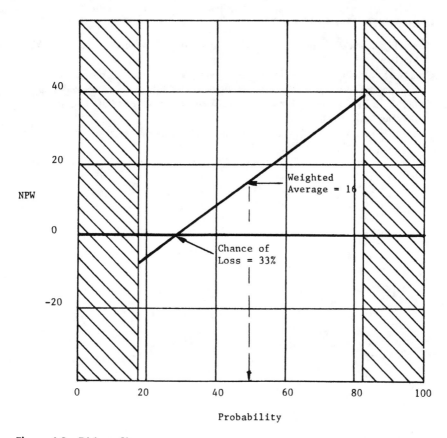

Figure 4.2 Risk profile.

The risk profile is plotted in Fig. 4.2. This chart plots the probability of producing a positive or negative cash flow. It is evident that the project has a little more than 30% chance that it will yield negative results. This situation is quite normal for most new product developments and does not necessarily mean that management should turn down the project; on the average, only one out of three new product attempts is successful.

REFERENCES

Cooper, D. O. and Davidson, L. B. (1976). The parameter method for risk analysis. *Chemical Engineering Progress*, Nov.:73–78.

Piekarski, J. A. (1984). Simplified risk analysis in project economics. *1984 AACE Transactions*, Morgantown, WV: AACE International, Paper D.5.

RECOMMENDED READING

Blank, L. T. and Tarquin, A. J. (2002). *Engineering Economy*. 5th ed. New York: McGraw-Hill Inc.

Canada, J. R., Sullivan, W. G. and White, J. A. (1996). *Capital Investment Analysis for Engineering and Management*. 2nd ed. Englewood Cliffs, NJ: Prentice-Hall Inc.

Collier, C. A. and Glagola, C. R. (1998). *Engineering Economic and Cost Analysis*. 3rd ed. Berkeley, CA: Peachpit Press.

Sullivan, W. G., Wicks, E. M. and Luxhoj, J. (2002). *Engineering Economy*. 12th ed. Englewood Ciffs, NJ: Prentice-Hall Inc.

Thorne, H. C. and Piekarski, J. A. (1995). *Techniques for Capital Expenditure Analysis*. New York: Marcel Dekker Inc.

5

Cost Control and Reporting

5.1 INTRODUCTION

AACE International has defined cost control (see Appendix B) as

> ... application of procedures to monitor expenditures and performance against progress of projects or manufacturing operations; to measure variance from authorized budgets and allow effective action to be taken to achieve minimum costs.

It is important to distinguish between cost control and cost accounting. Cost accounting is the historical reporting of disbursements and costs and expenditures; it is used to establish the precise and actual cost status of a project to date. Cost control, on the other hand, is used to predict the final outcome of a project based on up-to-date status.

According to Clark and Lorenzoni, a project cost control system has four objectives (1997, p. 146):

1. To focus management attention on potential cost trouble spots in time for corrective or cost-minimizing action (i.e., detect potential budget overruns before, rather than after, they occur).
2. To keep each project supervisor informed of the budget for each specific area of responsibility and how expenditure performance compares to budget.
3. To create a cost-conscious atmosphere so that all persons working on a project will be cost-conscious and aware of how their activities impact on the project cost.
4. To minimize project costs by looking at all activities from the viewpoint of cost reduction.

5.2   PROJECT FORMATS

There can be many different views of the objective of a cost control system depending on the owner, the contractor, or the designer. Construction projects are undertaken using different contract formats where the owner, contractor, and architect/engineer have different types of relationships with respect to the cost reporting and control functions. The Construction Industry Institute (CII Cost/Schedule Task Force, 1987) has identified four basic contracting formats:

> *Prime contractor* (or *general contractor, GC*). The contract for construction is between the owner and one contractor who has full responsibility for accomplishing the construction. The contract can be either fixed-price (lump sum, unit price, or combination thereof) or reimbursable. It can have special features such as a guaranteed maximum cost, and incentive features such as target cost or workhours with shared savings/penalties, milestone bonus/penalties, or other performance bonuses. The GC is responsible for establishing a cost control system, and the owner can be expected to require submission of a schedule of values and some agreed-upon mechanism for progress payments.
>
> *Multiple Primes managed by a construction manager* (*CM*). The CM function is handled by the owner or some agency under contract to the owner. The project is divided into major components or packages with separate contracts awarded for each component. The contracts may be between the owner and the contractors or between the CM and contractors. Normally the CM will establish the cost control system and require each contractor to operate within that system.
>
> *Turnkey.* The total management of engineering, procurement, and construction (EPC) is given to one firm, which may choose to handle all functions using its own resources or may choose to subcontract portions to others. The turnkey contractor will establish and maintain a cost control system.
>
> *Owner as general contractor.* The owner maintains a project management staff that directly manages company projects using a combination of direct-hire and subcontracting. The cost control system will be operated by the owner in this format.

5.2.1   Contracting Party Viewpoints

Each of the contracting parties in the project formats just listed is interested in cost control. The owner is interested in making sure that the project is completed within budget. The architect/engineer's involvement depends upon the level of design completion and the overlapping of engineering, procurement, and construction. The CM's interest is the same as that of the owner. The contractor and subcontractors are interested in detail cost control regardless of the contract type.

Project cost control requires cooperation between all parties in a project. Figure 5.1 shows the major information flow paths between the cost control system and other areas of concern.

5.3 ESTABLISHING BASELINES FOR CONTROL

The primary mission, or goal, of a project is the completion of a set of objectives that, when achieved, represents completion of the project. All contracting parties, whether the owner, the architect, the engineers, or the contractors, are interested in knowing how the project is coming along. To satisfy such an interest, a standard or baseline must be developed, against which progress may be measured. Identification of objectives is important, and should be done in the early stages of a project. Therefore, before data is collected or progress measured, a budget baseline is established to use as a yardstick to measure progress. Budget baselines are generated through the estimating process. If the project scope is not fully defined, this estimate will be approximate and may vary. As the scope becomes more defined, the budget estimates are updated.

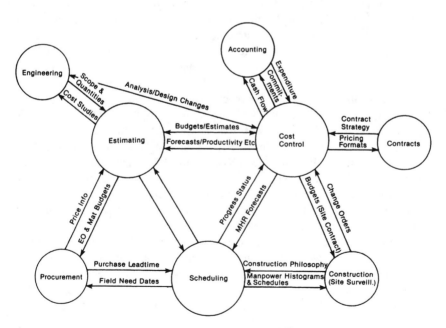

Figure 5.1 Information flow between cost control and other disciplines. (From K. Eckhoff and K. E. Harto. (1986). *Ninth International Cost Engineering Congress, Congress Papers,* Norwegian Association of Cost and Planning Engineering, International Cost Engineering Council, and Norwegian Petroleum Society, Oslo, N3.3, p. 2.)

5.3.1 Cost Control Variables

Budget baselines are determined by establishing various control categories. Establishing control categories that are fully compatible with the overall project objectives requires a review of four major cost control variables. These are (1) organization, (2) computerization, (3) level of control, and (4) elements of control (Stevenson, 1989).

Organization

Cost control during various stages of a project deals with variances between budgeted, earned, and actual costs in each control category. These categories are established based on an integrated control structure that takes into consideration what work is performed, who performs the work, who is accountable for the work, what the project's cost accounting code is, the project's contractual relationship, the project's work location, the project's system components, and so on—in other words, a work breakdown structure.

Work Breakdown Structure (WBS): A WBS is "a product-oriented family tree division of hardware, software, facilities and other items that organizes, defines and displays all of the work to be performed in accomplishing the project objectives" (AACE International Standard No. 10S-90; see Appendix B). A WBS can graphically display the work to be done, whether it is a division of engineering, procurement, or construction, and helps to correlate tasks, schedules, estimates, performance, and technical interfaces (Ahuja, 1980). The WBS acts as a vehicle for integrating baseline cost and time plan, and thus is an aid in relating plans to objectives.

The WBS also provides the mechanism for accumulation of actual and forecast costs and schedule data in support of overall project analysis and reporting. The WBS normally consists of two sections: definition and execution. The definition section is used to define the scope and establish initial cost estimates and schedules that will be used as a baseline against which actual and forecast information will be measured. The execution section defines the strategy selected for a particular project. The WBS is a matrix in which each item is referenced by two codes. First, a definition code defines the location and type of work being performed. Next, an execution code shows how these items will be packaged for contracting. Many projects fail to meet their specific objectives when project engineers develop a WBS that consists of only one section and that functions as "all things for all people." The proper procedure is to establish a WBS during the conceptual phase, expand it during the design phase, and then modify it as required throughout the life of a project.

WBS and Chart of Accounts: Breaking a project into many manageable subprojects using a WBS requires breaking down project cost along scheduled activities. Therefore a WBS should be accompanied by a *cost breakdown structure (CBS)* for cost planning. A CBS is a catalog of all cost elements expected within a project. The sum total of these elements will equal the project budget (Neil, 1988). The WBS is made up of project elements that are related to work tasks while the CBS is comprised of project elements that are related to cost control accounts. Figure 5.2 shows the relationship between WBS and CBS for an engineering-procurement-construction project (Neil, 1988).

Work Packaging: A work package is a well-defined scope of work that usually terminates once a deliverable product or service produces an identifiable or measurable result. A work package may consist of one or more activities or tasks with multiple resources and include one or more cost accounts (CII Cost/Schedule Task Force, 1988). These tasks may be related to such areas as engineering, procurement, manufacturing, construction, and start-up. For engineering, the deliverable tasks are the drawings, specifications, procedures, and

PHASES	INDIRECTS (1)	DIRECTS		
		Labor	Material	Equipment
Conceptual Engineering	$	WH $	$	$
Detailed Engineering	$	WH $	$	$
Procurement	$	WH $	$	$
Construction	$	WH $	$	$
Startup	$	WH $	$	$
Other (2)	$			

Legend: The COST BREAKDOWN STRUCTURE (CBS) is composed of all elements in the matrix for which dollars ($) are budgeted. The total dollar value of all of these elements equals the project budget.

The WORK BREAKDOWN STRUCTURE (WBS) is composed of those direct labor elements in the matrix for which work-hours (WH) are budgeted and lend themselves to work progress measurement.

Footnotes:
1. Supervision above first level, staff, facilities, supplies and services, travel, etc.
2. Home office overhead, contingency reserve, profit, etc.

Figure 5.2 Cost breakdown structure (CBS) and work breakdown structure (WBS). (From Neil, J. M. (1988). *Skills and Knowledge of Cost Engineering*, 2nd ed., J. M. Neil, ed., AACE International, Morgantown, WV.)

other items required by vendors or field personnel. For procurement, they include the delivery of materials and equipment. For manufacturing, the deliverable may be the whole product or a part of the product that is being produced. For construction, the deliverable is the constructed facility, or it may be a partial facility, such as an area or specific construction phase. For startup, the deliverables include system tests and operational certificates. A work package, therefore, provides the structure for efficient integration of engineering, procurement, manufacturing, construction, and startup.

Computerization

Although a manual approach to cost control may be appropriate for some small projects, today's computers are affordable for virtually any size job. Mainframes with work stations may be appropriate for very large projects, but personal computers (PCs) are now affordable for any project, no matter how small. There are many PC applications available that can be customized to meet a project's cost control needs. The needs of the project determine the type of system to use. As new software and hardware systems capture the market, technology constantly evolves. Advancements in computer technology have made it easy to network computer stations in order to share and transfer information between multiple users and to integrate various programs used for estimating, scheduling, work packaging, forecasting, and payroll, as well as other databases.

Mainframe computers use expensive hardware and software, but they provide for more centralized data processing and support. Mainframes also provide efficient remote job entry by allowing for better, faster, and more complete interaction between functions and locations on the project. PCs are used both as standalone processing units as well as acting as terminals transferring data and interacting between the mainframe and other work stations.

Level of Control

Determining the optimum level of detail to match the cost budget to the schedule is the key to developing the WBS. An optimum level of detail considers the project's size and complexity, the contracting philosophy of the parties involved, the project's organization and responsibility structure, and allows for a level of assumed risks and any operational constraints. At the corporate level, a project is generally viewed in a highly summarized form that focuses on the project's impact to the corporate budget, financial control of the project, and current reforecasting of project costs. At the project level, more detail is needed, although a summary form is still required. Cost control by contractors requires very detailed information about every task performed. A properly designed WBS can be summarized at the lowest level and be rolled up to provide the cost control information needed at the next higher level. An example of a WBS for a $25 million construction facility is shown in Fig. 5.3.

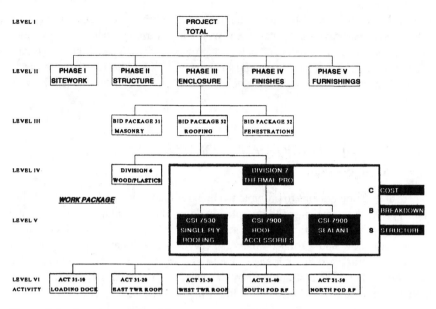

Figure 5.3 Sample work breakdown structure (WBS).

Elements of Control

A cost control system consists of five control elements that are tied to the various stages of a project (Fig. 5.4). The first step is to establish a baseline estimate to use as a plan to monitor and control costs. The estimate includes quantities, unit costs, productivity, wage rates, etc. The estimate is produced by the owner, the engineer, the construction manager, or with actual bids. "The very best estimate is some combination of all of these, helping to insure a more thorough understanding by project participants of the project's scope and cost expectations" (Stevenson, 1989). Once the baseline estimate is produced, monitoring the progress can begin. Data collection must be related to the same WBS used to establish the baseline estimate. Variance analysis is accomplished by comparing the actual

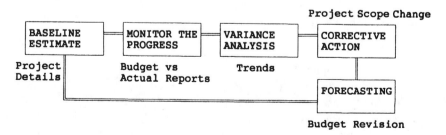

Figure 5.4 Elements of control.

data collected to the baseline estimate. Changes from the baseline estimate can be identified, thus providing management with early warnings that allow decisions and corrective actions to occur before a problem arises. The baseline estimate may require periodical reforecasting in order to include the results of variance analysis (Stevenson, 1989). These elements are discussed in detail in the following section.

5.4 STATUS EVALUATION AND CONTROL

5.4.1 Cost/Schedule Integration

There are several reasons to promote the integration of costs and schedules. Schedules produce the starting and completion dates, the percentage completed, and durations of the remaining tasks—essential information for a cost engineer calculating cash flow and variance analyses. The budget estimate provides data for workhour and craft breakdowns, quantities installed or produced, expended workhours and costs that are used by the scheduler to determine durations, percentages completed, labor force loading, earned workhours, and overall cost.

Performance Measurement

The performance measurement system ties together the various elements of a WBS, enabling an objective monitoring of project progress and performance via the concept of earned value (Mukho, 1982). In 1967 the U.S. Department of Defense (DOD) and the Department of Energy (DOE) established the Cost/Schedule Control Systems Criteria (C/SCSC) to use for control of selected federal projects. Although this system is primarily used on large projects, certain useful features of this system are applicable to other, smaller projects as well. The system consists of three major elements: the budgets, which are time phased to provide a *Budgeted Cost of Work Scheduled* (*BCWS*); the actual costs, which are captured as *Actual Cost of Work Performed* (*ACWP*); and the earned value concept, which is used to determine the *Budgeted Cost of Work Performed* (*BCWP*). By comparing these three major elements, several conclusions about cost and schedule performance can be made (Crean, 1982).

Measuring Work Progress

There are six methods for measuring work progress (Neil, 1988):

- Units completed
- Incremental milestone
- Start/finish
- Supervisor opinion
- Cost ratio
- Weighted or equivalent units

The *units completed method* can be used for tasks that involve repeated production of easily measured pieces of work. In the units completed method, a ratio is calculated by dividing the total units completed by the total units budgeted to generate a percent complete.

The *incremental milestone method* can be used for any control account that includes sequential subtasks. Completion of any subtasks is considered to be a milestone if it represents a specified percentage of the total installation when completed.

The *start/finish method* can be used for tasks that lack definable intermediate milestones, thus making it difficult to assign partial progress. With this method, an arbitrary percentage is assigned to the start of the task, and 100 percent is earned when the task is finished.

The *supervisor opinion method* is a highly subjective approach. This approach is used for tasks where a discrete method is not possible.

The *cost ratio method* can be used for tasks that either involve a long period of time or are continuous over the life of a project. In this method the percentage completed is calculated by dividing actual cost or workhours expended-to-date by those forecasted for completion.

The *method of weighted or equivalent units* is used for tasks that involve a long period of time and are composed of several subtasks, each with a different unit-of-work measurement. Each subtask is weighted according to the estimated level of effort required. As quantities are completed for each subtask, they are converted into equivalent units.

Earned Value or Achieved and Accomplished Values

The above methods are used for determining progress for a single type of work. *Earned value* or *achieved and accomplished value* are terms used for determining overall percentage completed of a combination of unlike work tasks or of a complete project. Earned value techniques are applicable to both fixed budgets and variable budgets, although the application of these techniques is different for the two situations. *Performance against schedule* is simply a comparison of what was planned to be done against what was done, while *performance against budget* is measured by comparing what was done to what has been paid for. The above can be expressed in these ways:

Performance Against Schedule

Scheduled Variance (SV) = BCWP − BCWS

Schedule Performance Index (SPI) = (BCWP)/(BCWS)

Performance Against Budget

Cost Variance (CV) = BCWP − ACWP

Cost Performance Index (CPI) = (BCWP)/(ACWP)

A positive variance and an index of 1.0 or greater reflects favorable performance. These calculations are used to determine forecasted costs for completion; there are three basic methods of using them:

Method 1 assumes that work from a specified point forward will progress at planned rates whether or not these rates have prevailed to this point. This is expressed as

$$EAC = ACWP + BAC - BCWP$$

where

   EAC = estimated at completion
   $ACWP$ = actual cost of work performed to date
   BAC = original budget at completion
   $BCWP$ = budgeted cost of work performed to date

Method 2 assumes that the rate of progress to date will continue to prevail and is expressed as

$$EAC = BAC/CPI$$

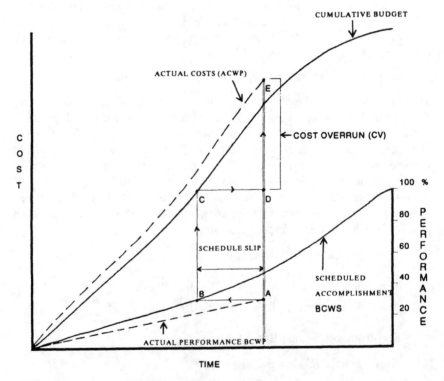

Figure 5.5   Earned value status graph (behind schedule and over budget).

where

    CPI = Cost Performance Index

    Method 3 uses progress curves for forecasting as shown in Figs. 5.5 and 5.6. Referring to Fig. 5.5, the actual accomplishment (point A) is plotted below the scheduled curve, indicating that the project is behind schedule. The actual amount can be determined by drawing or extending a horizontal line from point A back to point B on the schedule and then measuring the schedule slippage. Likewise, the plotted cost (point E) is located above the scheduled budget, but the amount of variation present in this parameter is not immediately apparent; the scheduled cost of the *actual accomplishment* must be determined rather than the cost listed for the *current time.* By extending a line vertically from Point B on the scheduled accomplishment until it meets the cumulative budget at Point C, we can determine what the cost for that accomplishment should have been. Continuing a horizontal line from that point over to Point D on the current time frame shows whether there is a cost overrun or underrun. In this case the cost overrun is measured as the vertical difference between D and E (Stevens,

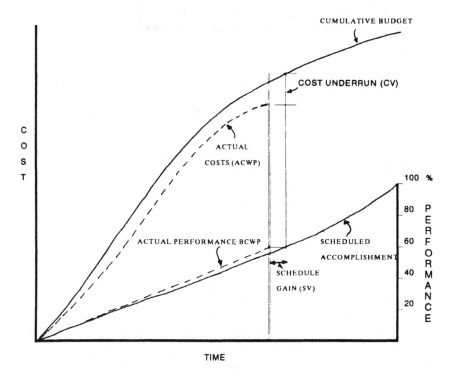

Figure 5.6 Earned value status graph (ahead of schedule and under budget).

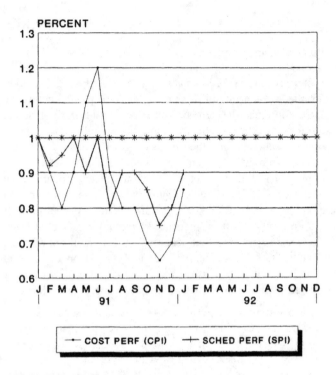

PERIOD ENDING: 12/31/91

Figure 5.7   Variance trends.

1983). Figure 5.6 shows a more positive case with the project both ahead of schedule and under budget. Other combinations are also possible.

   In addition to these curves, cost and schedule performance indexes can be plotted over time, thus providing variance trends as shown in Fig. 5.7. A good recommendation is that no single forecasting method be used. Using a forecast produced by each of the methods provides a range of possibilities and a more representative overview than any single method can provide.

5.5   REPORTING AND ANALYSIS

The purpose of cost control reports is to inform management of the project status. To be effective, the reports used for cost control should be brief, concise, timely, factual, and limited to pertinent information. The shorter and more concise the report, the faster feedback can be obtained and appropriate corrective action taken.

   Cost control reporting systems have progressed from manually-recorded information systems to today's computerized systems. Computerized systems can not only make extrapolations from status data, but they can also select information for inclusion in different reports. Computer-generated reports can also allow users to browse and view information on video displays. Simply having the information on a computer is not enough; appropriate information must be readily obtainable and presented in a clear, intelligible manner.

5.5.1 Frequency of Reports

Cost control reports should be coordinated with project scheduling as well as with accounting cutoff dates. Optimally these reports are issued each month (Patrascu, 1988). Cutoff dates for reports must be strictly adhered to in order to be of practical use; any major changes occurring after the cutoff date may be presented in an accompanying narrative report.

5.5.2 Variance Analysis

Variance reporting is a system that emphasizes cost and schedule control during all stages of design, construction, manufacturing, or operations. The purpose is to report to the client and the company all deviations from the budget. Every cost report should include an explanation of overruns and underruns since the last report. Furthermore, the cost engineer should be able to account for all changes made to the initial forecast at project inception.

5.5.3 Report Distribution

The best guide for the distribution of cost reports is to decide if the reports are needed for general information only or if they are actually used for decision making, and at what level those decisions are made. Depending on the intended distribution of the report, cost control reports can be grouped into the following categories:

Multilevel Cost Reporting. Not everyone in a company needs access to all of the information contained in the project control system. Multilevel reporting systems allow for different levels of objective determination and reporting. Examples of these include:

- Cost summary
- Labor rate
- Quantity and workhour
- Variance reporting.

Combined Cost/Schedule Reporting. These reports are generated by extrapolating information from both the cost and the schedule systems. Examples of these reports include:

- Cash flow report
- Cost/schedule performance curves
- Productivity profile
- Productivity trend chart
- Bulk quantity curves

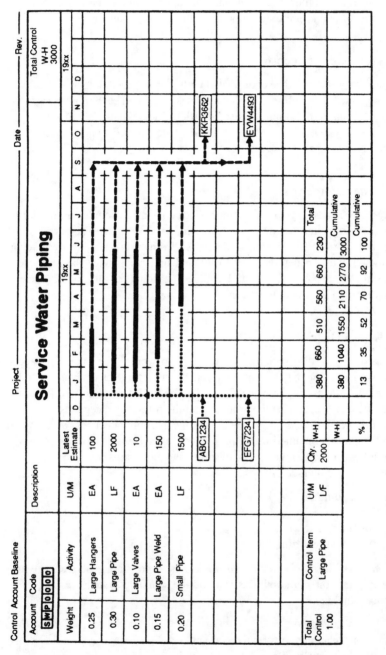

Figure 5.8 Control account for service water piping. (From Neil, J. M. (1988). *Skills and Knowledge of Cost Engineering*, 2nd ed., J. M. Neil, ed., AACE International, Morgantown, WV, p. 100.)

Control Account Baseline								Project ——————— Date ——————— Rev. ———————						

Account Code	Description												
S W P 0 0 0 0													This Period / To Date

Service Water Piping

Weight	Activity	U/M	Latest Estimate	Week Ending											
				1/6	1/13	1/20	1/27								
0.25	Large Hangers	EA	100	5 / 5	15 / 20	15 / 35	15 / 50								
0.30	Large Pipe	LF	2000				50 / 50								
0.10	Large Valves	EA	10												
0.15	Large Pipe Weld	EA	150												
0.20	Small Pipe	LF	1500												
Total Control 1.00	Control Item Large Pipe	U/M LF	Control Quantity 2000	25 / 25	75 / 100	75 / 175	90 / 265								

Field   Engineer

Figure 5.9 Monthly quantity report. (From Neil, J. M. (1988). *Skills and Knowledge of Cost Engineering*, 2nd ed., J. M. Neil, ed., AACE International, Morgantown, WV, p. 100.)

Miscellaneous Reports. A number of other reports may be products of a cost control system. Examples of these may include:

- Material requisition status
- Subcontractor status
- Vendor drawing status
- Fabrication status
- Critical items report
- Quality trends

Every project is a unique situation, each with its own specific reporting requirements. Some samples of common cost control reports are shown in Figs. 5.8 through 5.13.

To illustrate how a planner moves from the control to the detailed level, reference is made to Fig. 5.8. This figure is called a *control account baseline.* It is a document that takes a control schedule work package (in this case service water piping) and plans it out in detail. Note how the piping system is broken down into the work tasks required for its completion (large pipe, valves, etc). These are then scheduled in bar chart format. As is so often the case, these tasks are overlapping and there is some flexibility in their sequencing (soft logic). Use of the bar chart format with float shown for each bar gives the field the flexibility they need for accomplishing the work. Also note from the other information included on the baseline that the document provides the basis for earned value control and progress payments.

Figure 5.9 is a representation of a reporting format using the service water piping of Fig. 5.8 as an example. The many control accounts, in turn, can be summarized at various levels or for the whole project using earned value.

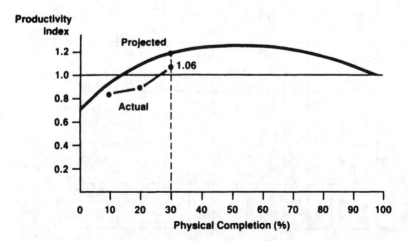

Figure 5.10 Productivity profile. (From Neil, J. M. (1988). *Skills and Knowledge of Cost Engineering,* 2nd ed., J. M. Neil, ed., AACE International, Morgantown, WV, p. 102.)

Figure 5.10 is an example graph for tracking productivity indexes. Note on this graph the use of a "projected" curve which does not coincide with the 1.0 datum line. This curve recognizes that productivity can usually be expected to be lower during the early stages of a project, reach a peak about midway in the project, and then decrease toward closeout. Keeping this in mind, the actual productivity plot can be more meaningfully evaluated. As shown on the example chart, a productivity index of 1.06, which one normally assumes is favorable, is actually low compared to what it should be for that point in time.

Figure 5.11 is an interesting variation of Fig. 5.10. Note how the vertical axis is workhours per percent complete. On this graph the cumulative plan curve is an upside-down image of the projected curve in Fig. 5.10 because of the different choice of units on the vertical axis. This graph also includes the plan for period and actual period plots to give it more usability. Note how the point identified as "(1)" shows that actual period performance equals that of planned performance. But, when you look at the actual cumulative performance, it shows that the project still has a problem because of the poor performance of prior periods and performance must become better than planned if the project is to recover.

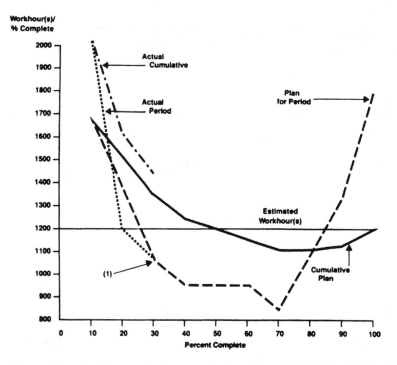

Figure 5.11 Workhour productivity trend chart. (From Neil, J. M. (1988). *Skills and Knowledge of Cost Engineering*, 2nd ed., J. M. Neil, ed., AACE International, Morgantown, WV, p. 102.)

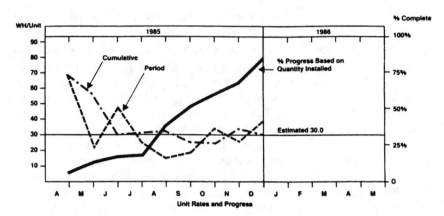

Figure 5.12 Building structural steel erection. (From Neil, J. M. (1988). *Skills and Knowledge of Cost Engineering*, 2nd ed., J. M. Neil, ed., AACE International, Morgantown, WV, p. 103.)

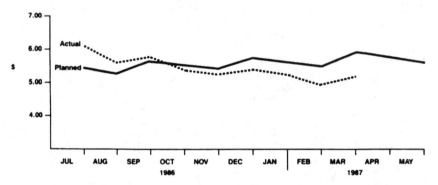

Figure 5.13 Unit wage rate. (From Neil, J. M. (1988). *Skills and Knowledge of Cost Engineering*, 2nd ed., J. M. Neil, ed., AACE International, Morgantown, WV, p. 103.)

Figures 5.12 and 5.13 track building steel erection workhour rates and unit wage rates, respectively, on a project and are self-explanatory.

REFERENCES

Ahuja, H. N. (1980). *Successful Construction Cost Control*. New York: John Wiley & Sons.

Clark, F. D., Lorenzoni, A. B. (1997). *Applied Cost Engineering*. 3rd ed. New York: Marcel Dekker, Inc.

CII Cost/Schedule Task Force (1987). *Project Control for Construction*. Austin, TX: Construction Industry Institute.

CII Cost/Schedule Task Force (1988). *Work Packaging for Project Control*. Austin, TX: Construction Industry Institute.

Crean, W. R. (1982). Applications of cost and schedule integration. *AACE Transactions*. Morgantown, WV: AACE International.

Mukho, S. (1982). Application of earned value for small project control. *AACE Transactions*. Morgantown, WV: AACE International.

Neil, J. M., ed. (1988). *Skills and Knowledge of Cost Engineering*. 2nd ed. Morgantown, WV: AACE International.

Patrascu, A. (1988). *Construction Cost Engineering Handbook*. New York: Marcel Dekker, Inc.

Stevens, W. M. (1983). *Cost Control: Integrated Cost/Schedule Performance*. Lockwood: Andrews & Newnam, Inc.

Stevenson, J. J. (1989). Cost control program to meet your needs. *AACE Transactions*. Morgantown, WV: AACE International, Paper F.1.

RECOMMENDED READING

AACE International. (2003). *Standard Cost Engineering Terminology*. AACE Standard No. 10S-90. Morgantown, WV: AACE International.

Bent, J. A. (1996). Humphreys, K. K., ed. *Effective Project Management through Applied Cost and Schedule Control*. New York: Marcel Dekker.

Hackney, J. W. (1997). Humphreys, K. K., ed. *Control and Management of Capital Projects*. 2nd ed. Morgantown, WV: AACE International.

Hermes, R. H. (1982). Cost-schedule integration: alternatives. *AACE Transactions*. Morgantown, WV: AACE International.

Humphreys, K. K., Wellman, P. (1996). *Basic Cost Engineering*. 3rd ed. New York: Marcel Dekker, Inc.

Nguyen, N. M. (1989). Cases impairing C/SCSC application in program management. *AACE Transactions*. Morgantown, WV: AACE International.

Riggs, L. S. (1988). Cost control: the industrial owner's view. *Cost Engineering*, 30(8).

Yates, J. K., Rahbar, F. F. (1990). Executive summary status report. *AACE Transactions*. Morgantown, WV: AACE International.

6

Project Management and Schedule Control

This chapter covers project management and schedule control for the engineering and construction of major capital projects. It covers project organization, conflict in execution of projects, planning, time scheduling, cost control, and project monitoring. It includes descriptions of the critical path method (CPM) and the precedence diagram method (PDM).

6.1 PROJECT MANAGEMENT, ORGANIZATION, AND EXECUTION CONFLICT

Figures 6.1, 6.2, and 6.3 show the three types of organizations: functional, projectized, and matrix. A functional organization (Fig. 6.1) is the standard pyramid type with the organization broken down into functional groups or departments, each reporting directly to a manager or executive. A projectized organization (Fig. 6.2) is, as the name suggests, organized by projects. A matrix organization (Fig. 6.3) is a combination of the other two types of organizations. In this arrangement, engineering, for example, reports to a manager or executive and also to various project managers. Engineering would thus have teams working on various projects.

Small projects and projects in the range of up to $100 million will generally use a matrix organization, while projects above $100 million will most frequently fall in the projectized organization category. The functional organization occurs chiefly where there are no engineering and construction capital projects, such as in a manufacturing operation.

In the most common matrix relationship, a general manager will be in charge of both a manager of projects and a function manager. (Depending upon the organization, vice-president, director, or some similar title may be

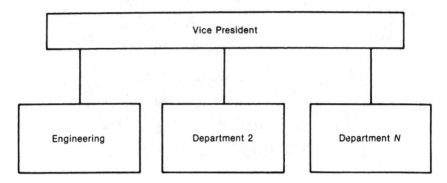

Figure 6.1 A functional organization.

used instead of manager.) Individual project managers will report to the manager of projects while all functional managers, such as engineering and accounting managers, will have project representatives reporting to them. The project managers and the function representatives then coordinate all project work, and assign subtasks as necessary. Figure 6.4 is a graphical representation of project control alternatives. During the course of major projects, organizations often tend to start with a functional management structure and evolve towards a project management structure, a left-to-right shift in this illustration. In the early stages, the majority of control and authority comes from the functional managers (central office), and is somewhat less than a matrix operation. As the project continues, the project manager (field office) increasingly picks up control and authority and the project becomes more of a matrix-type operation. Eventually the majority of the control rests in the field. The illustration shows a linear transition, but in practice the shape of this curve is decided by the

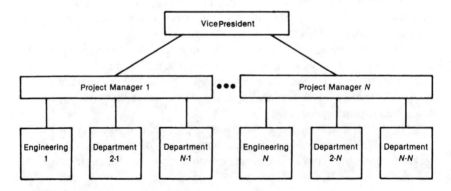

Figure 6.2 A projectized organization.

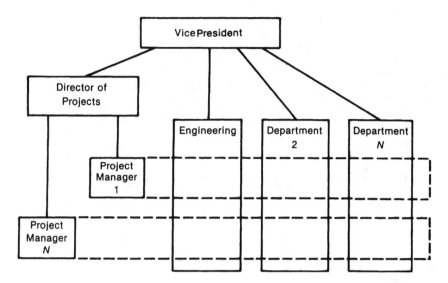

Figure 6.3   A matrix organization.

nature of the project and the mode of operation to which the particular contractor or business has become accustomed.

Of course, a matrix organization will produce conflicts since there are dual responsibilities for a number of facets of the project. Figure 6.5 demonstrates the overlap of responsibility for the budget. A project manager has responsibility for the budget. A project manager has responsibility for a project budget, which is the sum on budget figures horizontally across all organizational functions, as illustrated. Similarly, each function has a total budget that is distributed across multiple projects, the sum of any vertical column in Fig. 6.5. Frequently, as a result of this division of responsibilities, functional departments become

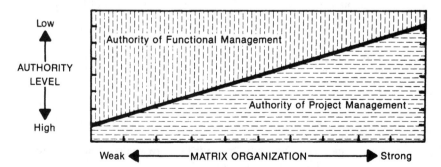

Figure 6.4   Spectrum of organizational alternatives. (From R. L. Kimmons, *Project Management Basics*, Marcel Dekker, Inc., New York, 1990, p. 21.)

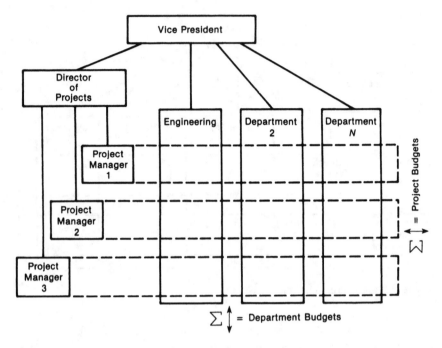

Figure 6.5 Budgets in the matrix organization.

overcommitted; this in turn promotes conflict and introduces politics to the project. When project managers lose control of the budget, they must bargain with the functional departments. Although a matrix organization normally gives the project manager the authority to requisition resources of the functional departments, the manager must be a diplomat to actually get the work done; there are other project managers competing for the same resources. Clearly the demands upon a department can exceed their ability to comply. Obviously, the budget requires careful planning. Consideration of the following items will contribute to the success of a matrix budget:

- Top management must define responsibilities and authority for the project manager.
- The project manager must anticipate conflicts.
- The project manager must take positive steps to develop teamwork.
- Documentation should be used to hold departments to their commitments.
- Functional managers should review and sign all documents relating to plans and schedules for the project.
- The project manager must avoid direct conflict with the department heads.

- The project manager should limit the task to "what" is to be done, not "how."
- The project manager must realize that each project is a new effort and that careful and continuous planning is necessary.

6.2 PROJECT MANAGEMENT AND CONTROL

Although many people have the ability to schedule, only a few are expert planners. The secret to a good plan is to put the whole project on a single sheet of paper, as shown in Fig. 6.6. This figure shows each phase of the project from the contract award to the mechanical completion; studying this figure will help develop an appreciation for the depth of information and timing presented on this single page.

Another example of a master plan is given in Fig. 6.7. It shows phase 1 and phase 2 shutdowns in the shaded areas. The shutdowns were important and critical, as they meant a loss of income to the refinery while the modernization was proceeding. A very simple plan for a small project is shown in Fig. 6.8 in the form of a bar chart.

The next step is to arrange a schedule from the plan. Key elements are the major milestones. Figures 6.6 through 6.8 show clearly such major milestones as contract award dates, mechanical completion dates, and the start of the shutdowns. The schedule must meet the major milestones to achieve the objectives set out in the plan. A first step in preparing the schedule from the plan is to prepare a plot plan, as shown in Fig. 6.9. A plot plan shows the times for deliveries of various pieces of equipment and such things as fabricated piping. Fig. 6.9 shows that the reactor steel is due in June. On large jobs a more elaborate plot is required, as shown in Fig. 6.10 for the fabricated steel. An overall plot plan shows multiple areas and work in blocks. It shows each area, the delivery dates, the vendors, and any other pertinent information. There will be individual plot plans for foundations, steel deliveries, piping deliveries, and so on. One has to know delivery dates to do a good scheduling job.

Equally important is the planning for labor, like the example shown in Fig. 6.11 for a moderate-size project. Labor requirements for a very large project are shown in Fig. 6.12. On large projects it can be difficult to satisfy the labor requirements. For this particular job good planning and scheduling had reduced the peak labor requirement from 3800 workhours to the shown level of 2500 workhours.

The end result of all planning is to come up with a time-phased plan; this process is called *scheduling*. Figure 6.13 is a simple maintenance schedule shown as a bar chart or Gantt chart (see following discussion).

Figure 6.14 is a procurement schedule. It shows an equipment list and everything that must be done for the equipment at each of the various milestones.

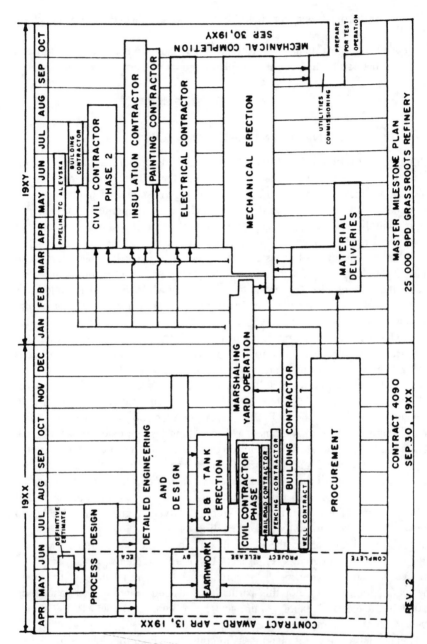

Figure 6.6 Master milestone plan.

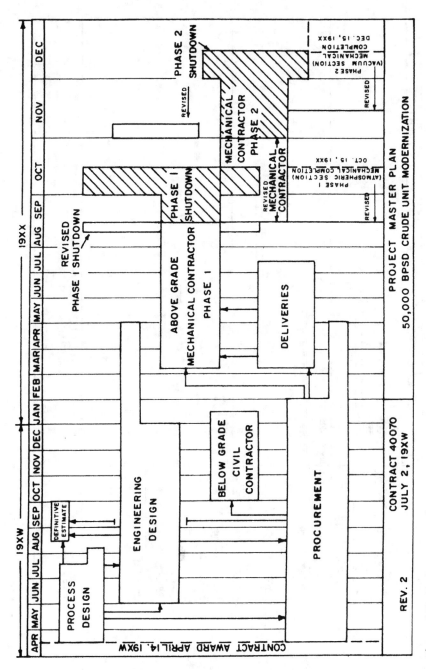

Figure 6.7 Project master plan.

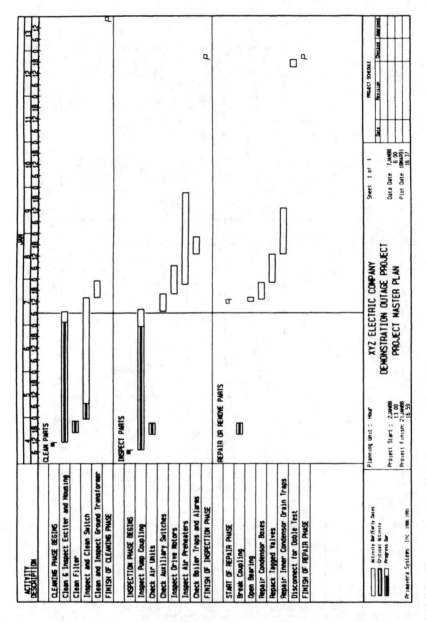

Figure 6.8   Bar chart for project master plan.

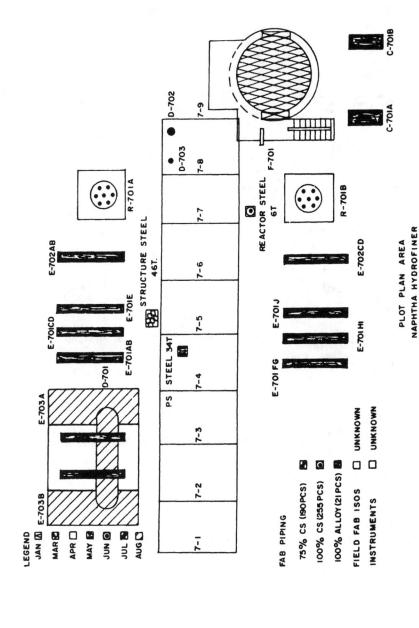

Figure 6.9 Plot plan.

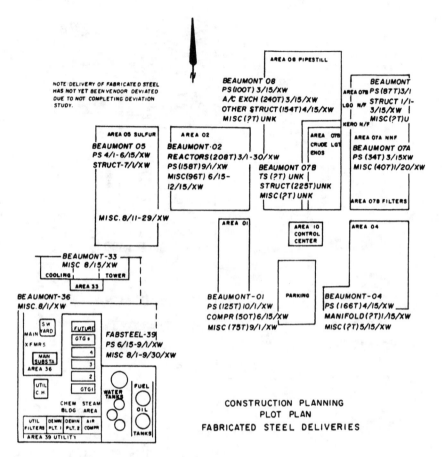

Figure 6.10   Plot plan—fabricated steel deliveries.

The engineering status shows each of the milestone dates, such as process data and issue of specifications. Procurement is timed by scheduling each major milestone for each major piece of equipment. This can be an involved effort; this task should be computerized on large projects which may have more than 2000 pieces of major equipment. In practice, scheduling of this nature is generally computerized, even on small projects. Excellent PC-based software is available to facilitate the scheduling function.

Figure 6.15 is a construction schedule. It is a weighted bar chart and the triangles show original delivery dates, revised delivery dates, and the actual arrival. This is a popular chart format for use in the field, easily understandable with a quick glance.

Another example of a construction schedule, but one including workhours, is shown as a bar chart in Fig. 6.16. There is an interdependency here since the

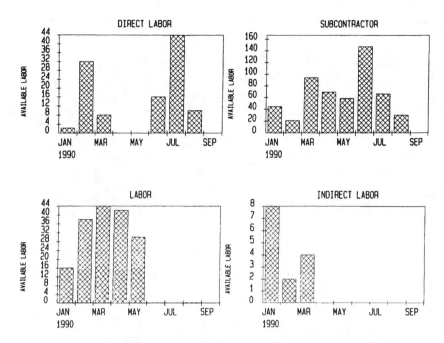

Figure 6.11 Anticipated construction labor.

steel must be placed and a vessel must be set before the piping effort can begin (see following discussions on CPM and PERT).

These figures all show basic types of scheduling that will work for engineering, procurement, and construction. Next, we have to monitor the project itself to ensure that the schedule is maintained. Figures 6.17 through 6.20 are all examples of progress reports for engineering and procurement. (Later, progress reports for construction will be shown.) Figure 6.17 is an office schedule performance summary which shows percent completion for the various tasks in an engineering office and the total for the project. It is an interim report that shows what percent was completed on a particular date, the percent completed 2 weeks earlier, and what was actually accomplished within the 2-week period.

Figure 6.18 shows curves for process and flowsheets as well as for estimating and cost. Similar curves are developed for project engineering and for procurement. All this needs to be augmented with computer printouts like those shown in Fig. 6.19, a status report for work done at a refinery, and Fig. 6.20 for another refinery project.

At this point we have covered progress monitoring for engineering and procurement of a contractor or an operating company which has a department for engineering and procurement. The next consideration is

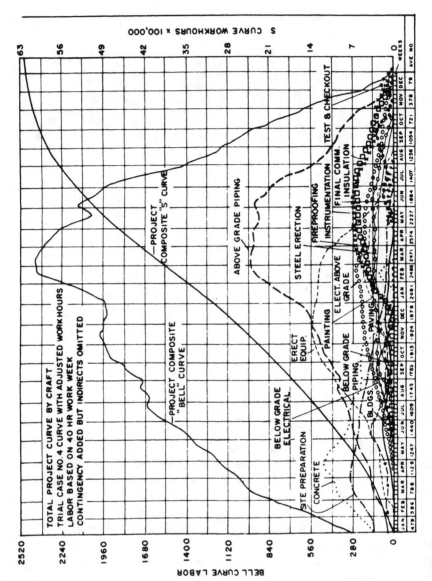

Figure 6.12 Labor requirements for large project.

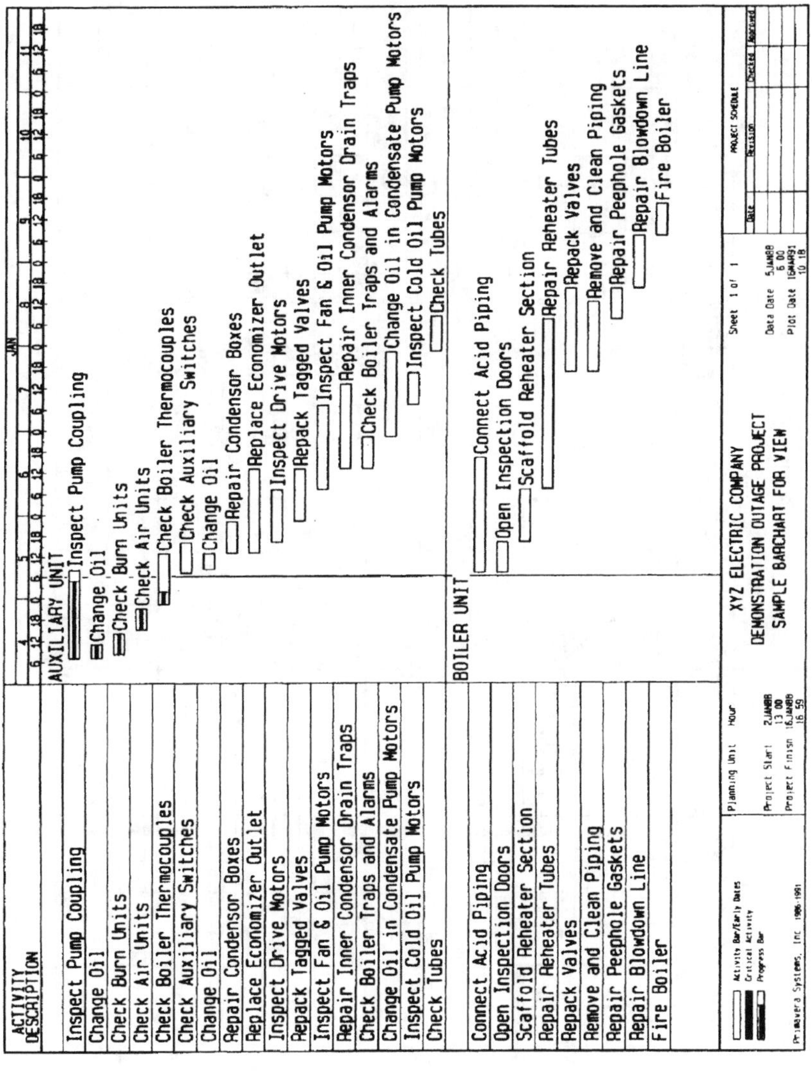

Figure 6.13 Bar chart for small nonconstruction project.

MASTER EQUIPMENT LIST

CONTRACT 4109 | UNIT 25.1
CLIENT
COST CODE 1110 | CATEGORY Towers
BY STEVE | APPO
DATE 8-12-XX | PG. 1 OF 10 | REV 1

ITEM NO. / COST CODE	DESCRIPTION AND SIZE	WEIGHT LBS	REQN NO	REQN REV / DATE	P O NO REV / DATE	VENDOR / REMARKS
95-201 / 1111-01	CRUDE TOWER 19'-0" I.D. x 148'-1" T/T 192'-0" OAL	651,850	1111-1	0; 1	0; 1	A285C w/Monel & 410s Clad
1112-01	CRUDE TOWER TRAYS 48		1112-1	0 6/21; 1	0; 1	12 Cr
95-202 / 1111-02	HEAVY NAPHTHA STRIPPER 5'-0" I.D. x 23'-6" T/T	63,675	1111-2	0 6/29; 1 7/22	0; 1	Stacked Tower 91'-8" O.A.L. A285C
1112-02	HEAVY NAPHTHA STRIPPER TRAYS 6		1112-2	0 6/29; 1	0; 1	12 Cr.
95-203 / 1111-02	NO 2 FUEL OIL STRIPPER 9'-6" I.D. x 24'-0" T/T		1111-2	0 6/29; 1 7/22	0; 1	A285C w/410s Clad
1112-02	NO.2 FUEL OIL STRIPPER TRAYS 6		1112-2	0 6/29; 1	0; 1	12 Cr.
95-204 / 1111-02	LGO STRIPPER 5'-0" I.D. x 31'-6" T/T		1111-2	0 6/29; 1 7/22	0; 1	A285C w/410s Clad
1112-02	LGO STRIPPER TRAYS 6		1112-2	0 6/29; 1	0; 1	12 Cr
95-205 / 1111-03	DEPENTANIZER 7'-6" x 9'-6" I.D. x 146'-6" T/T O.A.L.	310,170		0; 1	0; 1	
1112-03	DEPENTANIZER TRAYS 64			0; 1	0; 1	12 Cr
95-206 / 1111-04	AMINE CONTRACTOR 6'-0" I.D. x 48'-6" T/T 56'-9" O.A.L.			0; 1	0; 1	A515GR70
1112-04	CONTRACTOR PACKING & INTERNALS			0; 1	0; 1	

EQUIPMENT STATUS

LEGEND: L = LETTER; S = SLIPPAGE; T = TELEPHONE; V = VISIT

SHIPMENT CODE: UP TO 50 %; 50 % - 75 %; 100 %; RECEIVED; CANCELLED; HOLDING

DATE 8-12-XX | PG 1 | OF 10 | REV 1

Row labels per item: REQD / ORIGINAL / CURNT S / CKD BY; ACTUAL SHIPPING DATES: SHIPMT CODE / FIRST / LAST; DATES: REQD at SITE / VIA / ETA at SITE

ITEM NO.	SCHEDULED DATES: DWGS FOR APPR (REQD)	DWGS CERT (REQD)	SHIP FROM SHOP
95-201	9-1	2-1	
1112-01	9-9	12-1	
95-202	10-12	12-15	
1112-02	9-15	12-15	
95-203	10-12	12-15	
1112-02	9-15	12-15	
95-204	10-12	12-15	
1112-02	9-15	12-15	
95-205	10-14	12-15	
1112-03	10-20	12-3	
95-206	10-10	12-31	
1112-04			

ENGINEERING STATUS

DATE 8-12-XX | PG 1 | OF 10 | REV. 1

ITEM NO.		PROCESS DATA	ISSUE SPEC. & REQN	INQUIRY ISSUED	QUOTES REC'D	BID TAB	CLIENT APP'D BID TAB	RELEASE FOR P.O.
95-201	SCH	4-27	5-25	5-26	6-24	7-20	7-29	8-1
	ACT	4-27	5-25	5-26				
				INQ. NO 1-1100				
1112-01	SCH	6-10	6-21	6-22	7-19	7-27	8-10	8-11
	ACT	6-10	6-21					
				INQ. NO 2-1100				
95-202	SCH	7-15	7-21	7-22	8-19	8-26	9-5	9-9
	ACT	7-16	7-23	7-25				
				INQ. NO 4-1100				
1112-02	SCH	6-24	6-24	6-29	7-22	7-29	8-11	9-12
	ACT	6-20	6-30	6-30				
				INQ. NO 5-1100				
95-203	SCH							
	ACT							
		SEE H.N. STRIPPER		INQ. NO				
1112-02	SCH							
	ACT							
		SEE H.N STRIPPER TRAYS		INQ. NO				
95-204	SCH							
	ACT							
		SEE H.N. STRIPPER		INQ. NO				
1112-02	SCH							
	ACT							
		SEE H.N. STRIPPER TRAYS		INQ. NO				
95-205	SCH	7-12	7-27	7-28	8-26	8-31	9-11	9-12
	ACT	7-12						
				INQ. NO				
1112-03	SCH	7-15	7-27	7-28	8-26	9-1	9-15	9-16
	ACT	7-15						
				INQ. NO				
95-206	SCH	7-12	7-27	7-28	8-26	9-1	9-15	9-16
	ACT	7-12						
				INQ. NO				
1112-04	SCH							
	ACT							
				INQ. NO				

Figure 6.14 Procurement schedule.

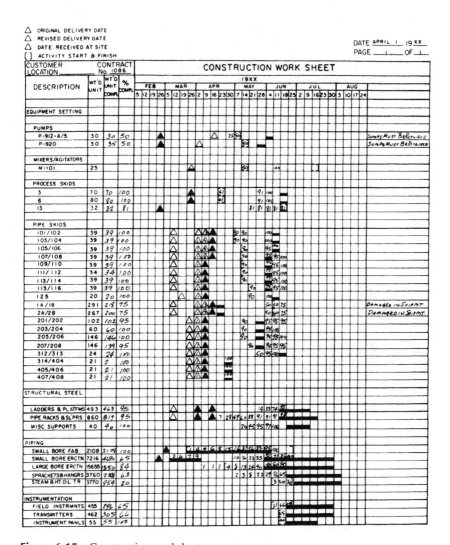

Figure 6.15 Construction worksheet.

construction cost monitoring. The principles remain the same, and Fig. 6.21 through 6.23 illustrate various ways to display progress. Figure 6.23 features a curve for the field indirect costs, which are increasingly becoming a major part of large jobs.

The whole purpose of reporting is to avoid surprises and to permit taking corrective action early enough to prevent a major deviation. Good field cost reporting is mandatory on a weekly basis; the weekly report should be on the construction manager's desk by noon on Monday.

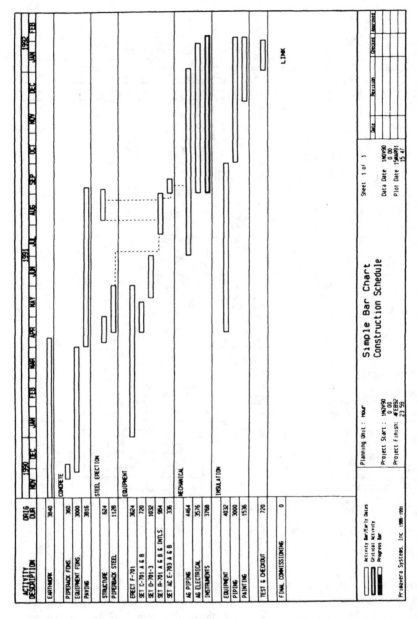

Figure 6.16   Simple bar chart construction schedule.

DESCRIPTION	PROGRESS (% COMPLETE)			SCHEDULE			WORKHOURS		
	As of 9/25	As of 9/11	2-wk Gain	% Sched	Ahead+ Behind-	2 Wks Ago	% Wkhr	Under- Over+	2 Wks Ago
Project/Equipment	50.5	48.5	2	50	+0.5	+1.5	50	-0.5	-1
Process/Flow Sheets	91.5	90.5	1	94	-2.5	-0.5	97.5	+6	+4
Purchasing	61.5	54.5	7	47	+14.5	+10.5	42.5	-19	-15.5
MTO & Estimating	79	76.5	2.5	62	+17	+18.5	47.5	-31.5	-31.5
Inspection/Expediting	24	21	3	19	+5	+6	29	+5	+2
Cost & Scheduling	24.5	23.5	1	22	+2.5	+2.5	14.5	-10	-10
Design	85	79	6	82	+3	+5	77	-8	-9
Vessels	91	89	2	88	+3	+7	66.5	-24.5	-25
Civil/Structural	98	98	0	87	+11	+17	68	-30	-31.5
Piping	74.5	73.5	6	81	-1.5	+1/2	83	+3.5	+1
Electrical	69	55.5	13.5	70	-1	-6.5	80	+11	+9
Instruments	86.5	76.5	10	78	+8.5	+6.5	76	-10.5	-8.5
TOTAL PROJECT	63	59	4	58	+5	+5	55.5	-7.5	-8

Figure 6.17 Office schedule performance summary.

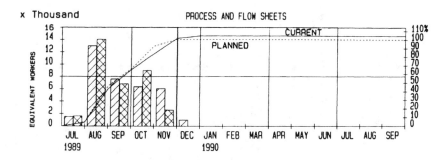

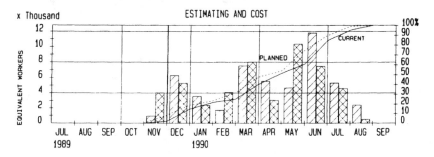

Figure 6.18 Schedule progress curves—plant expansion and modernization.

STATUS REPORT
DATE 08/16/XX
ENT380301

CUSTOMER
UNIT BTX EXP + AROMATICS SPLTR
LOCATION

REPORT NBR 6
PER / END 08/13/XX

CONTRACT 041.08
UNIT NBR 00/0
OPER NBR 22

SCOPE CHANGE ITEM / DWG NBR	DESCRIPTION	SCHEDULED DATES START----COMPLETE	ACTUAL DATES START----COMPLETE	EST HOURS	PCT COMP	ISSUE REV----DATE	STATUS ----DATE	NOTE	NOTE CODE
A01	SUPERVISION, MISCELLANEOUS VENDOR DWG. LOGGING	04/20/XX 10/30/XX	04/20/XX	280	85	CUR- AFC-			
A02	RAFFINATE WATER WASH COLUMN V-15	04/25/XX 07/15/XX	04/25/XX 07/27/XX	20	100	CUR- AFC-			
A03	EXTRACTOR WATER WASH COLUMN V-42	04/25/XX 07/15/XX	04/25/XX 07/27/XX	20	100	CUR- AFC-			
BTX-V-100	REWORK OF EXISTING BENZENE TOWER ITEM V-1	04/20/XX 07/30/XX	04/26/XX 06/29/XX	40	100	CUR- 2 AFC-	06/29/XX	THIS DWG IS NOT FOR CONSTRUCTION	
BTX-V-100V1	VENDOR CHECK (NOZZLES)	06/15/XX 07/30/XX		4		CUR- AFC-		LATE STARTING LATE COMPLETING ISSD FOR QUOTE	502
BTX-V-100V2	REWORK OF EXISTING BENZENE TOWER V-1 VENDOR CHECK (TRAYS)	06/15/XX 07/30/XX	07/18/XX	8	90	CUR- AFC-		LATE COMPLETING	
BTX-V-101	REWORK OF EXISTING TOLUENE COLUMN ITEM V-3	06/22/XX 08/15/XX	06/15/XX 08/11/XX	20	100	CUR- 0 AFC-	06/29/XX	THIS DWG IS NOT FOR CONSTRUCTION	
BTX-V-101VC	REWORK OF EXISTING TOLUENE COLUMN VENDOR CHECK (VESSEL) V-3	07/15/XX 08/15/XX	08/11/XX	8	100	CUR- AFC-			
BTX-V-102	REWORK OF EXISTING EXTRACTOR ITEM V-13	06/06/XX 08/15/XX	06/23/XX 08/11/XX	25	100	CUR- 0 AFC-	06/29/XX	THIS DWG IS NOT FOR CONSTRUCTION	
BTX-V-102V1	REWORK OF EXIST. EXTRACTOR V-13 VENDOR CHECK (NOZZLES)	07/01/XX 08/15/XX	08/11/XX	8	100	CUR- AFC-			
BTX-V-102V2	REWORK OF EXIST. EXTRACTOR V-13 VENDOR CHECK (TRAYS)	07/15/XX 08/15/XX	05/25/XX	30	30	CUR- AFC-		DUE TO COMPLETE	
BTX-V-103	MISCELLANEOUS DETAILS OF EXIST. EXTRACTOR ITEM V-13	06/06/XX 08/15/XX	06/23/XX 08/11/XX	15	100	CUR- 0 AFC-	06/29/XX	THIS DWG IS NOT FOR CONSTRUCTION	

REPORT NBR 6
PAGE NBR 11

Figure 6.19 Status report.

PROJECT STATUS SUMMARY
DATE 09/27/XX
EN7380302

CUSTOMER
PROJECT NO 3A PIPE STILL EXP - GOF
LOCATION

REPORT NBR 6
PER / END 09/24/XX

CONTRACT: 4119

OPER NBR	OPERATION DESCRIPTION	PCT COMP	SCHED PCT COMP	+AHEAD -BEHIND SCHED	PCT GAIN THIS PER	CURR HR	TOTAL HR TO DATE	FRCST TO GO	TOTAL HR FRCST	BUDGET HR	PCT BGT EXPND
21	FLOT PLANS	90	80	+10	0	0	14	116	130	100	14
22	VESSEL ENGINEERING	39	39		9	84	329	61	390	390	94
23	PIPING ENGINEERING	1	10	-9	9	6	27	335	362	350	18
24	PIPING DESIGN	10	10		5	180	360	1642	2002	2030	18
25	PIPING ISOMETRICS	0	0		0	0	0	1870	1870	2020	0
26	MATL CONTROL PIPING	2	1	+1	0	0	16	398	414	410	4
27	CIVIL/STRUCT ENG	34	34		13	156	364	389	750	560	65
28	CIVIL/STRUCT DESIGN	6	10	-4	2	24	85	1270	1355	1050	8
31	ELECTRICAL ENGINEERING	21	21		2	18	107	499	606	410	26
32	ELECTRICAL DESIGN	6	6		4	54	85	615	700	710	12
33	MATL CONTROL - ELEC										
34	INSTRUMENT ENGINEERING	12	12		7	115	214	1135	1349	880	24
35	INSTRUMENT DRAFTING	0	2	-2	0	17	36	754	790	490	7
37	PLANT MODELS										
--	REQUISITION REPORT										
	TOTAL ENGINEERING OPERATIONS	10-1/2	10-1/2			654	1637	9081	10718	9400	17
10	ESTIMATING	0				0	0			0	0
11	COST CONTROL	27	27		4	12	78	357	435	250	31
12	SCHEDULING	32	32		7	32	145	125	270	280	52
13	PURCHASING	11	11		6	32	56	504	560	560	10
14	EXPEDITING	0	2	-2	0	21	28	462	490	490	6
15	INSPECTION	6	6		0	38	79	361	440	440	18
16	TRAFFIC										
17	VENDOR PRINT CONTROL										
18	SECR + CL - REIMB	22	18	+4	8	0	1	589	590	0	0
19	SECR + CL - NONREIMB					5	55	1295	1350	2395	2
20	FLOW SHEETS	78	78		19	65	94	0	94	0	0
39	FLD CK+ST-UP NONREIMB					22	195	40	235	130	150
40	SECTION CHIEFS	0				0		0	0	0	0
41	ENGINEERING MANAGEMENT					1	23	0	23	0	0
42	FLD CHECKOUT - REIMB						-6		-6		
53	PROJECT ENGINEERING	31	31		5	80	429	1088	1516	1460	29
54	PROJECT MANAGEMENT	22	21		3	40	389	817	1206	2270	17
56	GENERAL PROCESS	72	72	+1	13	22	433	32	465	180	241
57	OPERATING MANUALS										
58	FLD ST-UP - REIMB										
	TOTAL PROJECT OPERATIONS					370	2010	5730	7740	9045	22
	TOTAL PROJECT					1624	3647	14811	18458	18445	20

Figure 6.20 Project status summary.

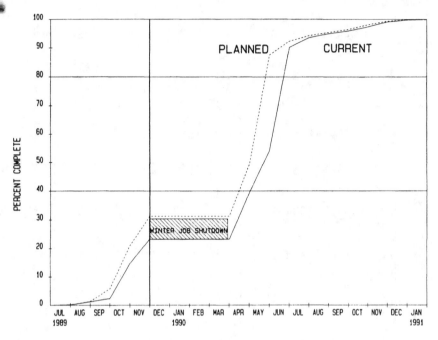

Figure 6.21 Total construction progress curve.

6.3 PERT, CPM, ARROW AND PRECEDENCE SCHEDULING

The suspected root cause of project failures is frequently attributed to "poor planning." AACE International defines planning as

> ...the determination of a project's objectives with identification of the activities to be performed, methods and resources to be used in accomplishing the tasks, assignment of responsibility and accountability, and establishment of an integrated plan to achieve completion as required.

A project may be defined as any effort undertaken to accomplish some specific objective within a given time and budget restraints. Therefore, depending on the scope of a project, the required planning effort will vary significantly. The range of efforts may vary. It could be a one-page scope, schedule, or budget statement for a very simple project, or it could be a full project-control effort that involves an integrated computerized cost/schedule control system and a scope-of-work document. Such a complex document could include concomitant work and organizational breakdown structures for either a complex multicontractor development program or for a design and construction plan of a large manufacturing facility.

Scheduling, the establishment of specific activity dates within the constraints of the plan, was for many years done graphically by use of Gantt

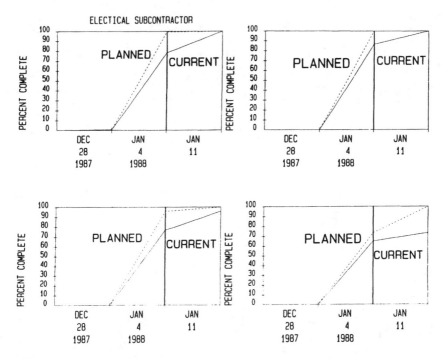

Figure 6.22 Electrical progress curves.

charts as shown in Fig. 6.13. This time-scaled display of activities plotted as bars over a time line (with progress shown by varying the bar appearance) provides an excellent graphical portrayal for working schedules; it is still emulated by many current scheduling software programs. The primary difficulty with the Gantt chart is that the underlying logic for the plan as developed by the project managers and schedulers is in their minds at the time of schedule development and frequently is not documented. This situation makes it difficult to perform an assessment of overall schedule impacts. This was the status of general scheduling until the 1950s when the Critical Path Method (CPM) was developed. Also, in the early 1950s the programmed evaluation and review technique (PERT) was developed for the U.S. Navy's Polaris missle program; it has since been modified to include cost.

CPM involves the development of networks that include both the identified activities and their logical dependencies. The development of networks has been approached in two formats: the arrow diagram method (ADM), and the precedence diagram method (PDM). ADM is called activity-on-arrow while PDM is called activity-on-node. Figures 6.24 and 6.25 show the basic differences graphically.

Project planning begins with a well-defined scope of work, a set of objectives, and an assigned management team. Although the scope will likely

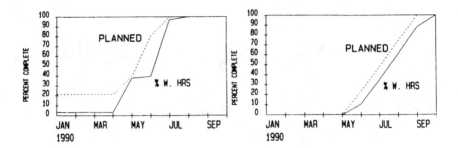

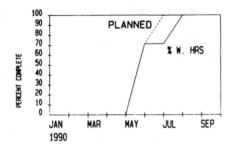

Figure 6.23   Indirect costs.

evolve as the project continues, it is best to document the initial basis of a project plan to serve as a reference point for future adjustments. The assigned team should then develop a work breakdown structure (WBS) that will serve as the focal point for future control of the project.

The WBS is a product-oriented tree diagram that graphically portrays the project scope in terms of manageable blocks of work. Each level of the WBS details those products or deliverables that are required to support the next or higher level. With the WBS established, the team can select the level at which it intends to manage or control the project and can then begin the establishment of budgets and schedules for each work package.

Each work package should have the following characteristics:

- The scope should be clearly defined.
- It should have a budget.
- It should be assignable to a single organizational element.
- It should be clearly distinguished from other work packages.
- It should have distinct start and completion dates.
- It should be possible to objectively measure progress.

Typically, during the initial stages of a project before development of the WBS, a master schedule or summary schedule would be developed for

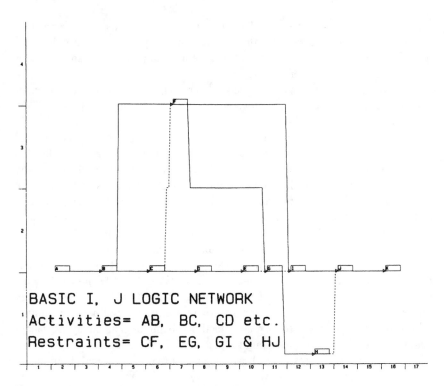

Figure 6.24 Arrow diagramming method (ADM).

management review. This schedule could be developed by the project team without the benefit of a totally integrated plan. As such, it would require significant expertise and understanding of the type of work being undertaken. The project planner would have to work closely with the team in developing the approximate blocks of time required for the significant phases of the project, whether it be engineering, procurement, or installation. The planner and the team would also have to work jointly to define the acceptable overlaps in these areas. During team review and discussion, a block diagram such as

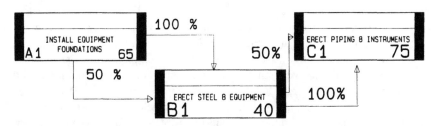

Figure 6.25 Precedence diagramming method (PDM).

that previously shown in Fig. 6.7 might be developed to assist in presenting or visualizing critical aspects of the project, such as preestablished milestones or outages. This block diagram might then be used in preparing Gantt charts or milestone lists for management presentation.

With the project scope packaged, the project control engineer or planner and the project team are now responsible for developing the project CPM network. This process, although fairly straightforward, can vary significantly with the scope, the number of organizations involved, and the uniqueness and complexity of the project being undertaken. The planner and the team develop the logical relationships between the work packages, being careful to use only definite, precise relationships to assure maintaining maximum flexibility in the plan. It is the planner's specific responsibility to question the team carefully and to ensure that relationships are not merely preferences but are indeed physical constraints. Figure 6.24 depicts a pure logic diagram in ADM format with events (or nodes) shown as rectangles; for example, A is the start, intermediate events are B, C, ..., J, and the finish is K. The arrows between events are called *activities*. The dotted arrows between C and F, E and G, G and I, and H and J are logical restraints, sometimes called "dummy activities." They consume no time, but they reflect interdependence between activities and are necessary for establishing a proper sequence in the network. Any activity beginning at an event cannot begin until all the activities leading to that event are completed.

The team then assesses each activity to estimate an elapsed duration for completing each package of work. Activity descriptions are also added to describe what is being physically accomplished within the work package. They are usually brief and should be limited to a descriptive action verb/object combination such as "prepare boiler specification." In Fig. 6.26, the same network is shown with activity descriptions and durations. The abbreviations used on the figure are LB for large-bore piping, SB for small-bore piping, and H/T for hydrostatic testing. Activity A-B is to erect 25% of the large-bore piping in 25 days. This brief activity description should be supplemented within the work package scope, or a supplemental punchlist should be provided to delineate the specific areas necessary to complete the work package. Restraints are shown with no duration and there is a milestone shown for mechanical completion at the termination of activity J-K.

Figure 6.27 lists the original duration for each activity. This is the time for the planners to begin calculation of the project *critical path*—the longest calculated path duration between the start and finish of the project's activities. The first step in this calculation is called the *forward-pass* calculation, the determination of the earliest dates activities can be started and completed. As the planners progress through the network, they take an activity's original duration and add to it the early finish of a preceding activity, thereby determining its earliest possible finish. At each junction point the planner must select the latest finish of all incoming activities to determine the earliest possible start for succeeding activities. Looking at activity F-I (in Fig. 6.27), the planner must decide what its early

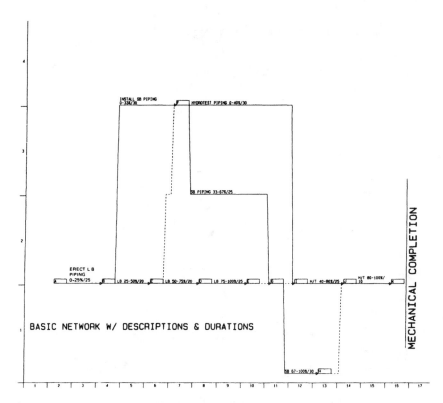

Figure 6.26 CPM example with activity descriptions and durations.

start is based on the completion of prior activity logic. Activity string A-B, B-C, C-F (logical restraint) would be complete by day 45 (25 + 20 + 0), while string A-B, B-F would be complete by day 55 (25 + 30). Therefore the earliest that activity F-I could start would be on day 56. When calculating the forward pass, the latest arrival date of all incoming activities must be used. Figure 6.28 includes a tabulation of early starts and finishes for the network activities and shows that the earliest possible completion date for this project is day 130.

The next step in the critical path determination is the *backward-pass* calculation, in which the latest possible start and finish dates are determined for each activity (see Fig. 6.29). The calculations are made beginning at project completion (event K) and working backward through the network. At events where there are two or more activities going forward, the earlier date is chosen. At event G, the latest possible day activity G-H could start is K minus the activity durations for J-K, H-J (logical restraint), and G-H, which turns out to be day 90 (130-10-0-30). The latest day activity G-I (logical restraint) could begin is K minus the activity durations for J-K, I-J, and G-I, or day 95 (130-10-25-0). Therefore, if we're not going to impact project completion at day 130, we

START OF BASIC NETWORK LISTING

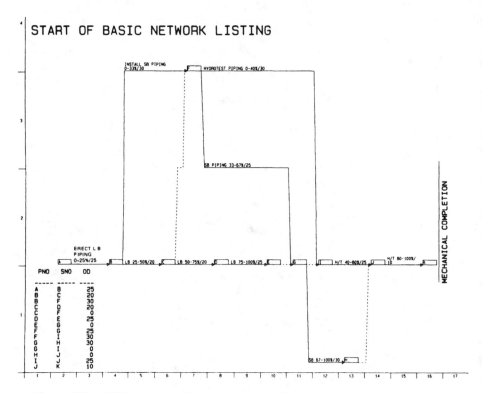

Figure 6.27   CPM example with original duration listing.

must complete any activity coming into event G by the earlier of the two dates, day 90. Then activity F-G would have to start no later than day 66 to support completion of the project by day 130. Figure 6.29 tabulates the late start/finish dates from the backward pass calculation, along with the early dates.

In comparing the early and late start dates for an activity such as B-F, the difference of 10 days is considered *total float*. Although B-F could finish as early as day 55, delaying its completion to day 65 would not impact project completion by day 130 since activity F-G can start as late as day 66. The tables on Fig. 6.29 and 6.30 tabulate the schedule comparison and total float for each activity. Activities having zero total float must be finished within the allotted duration and are considered to be on the *critical path* of the network, because delays in their completion will have a direct impact on project completion by day 130 unless alternative actions are taken. Note that if the required completion of the project were day 150, the project would have 20 days of total float; however, the critical path would remain through activities A-B, B-C, C-D, D-E, E-G, G-H, and J-K. There are other paths, but they are subcritical because they have some total float at the outset of the project. These subcritical activities must

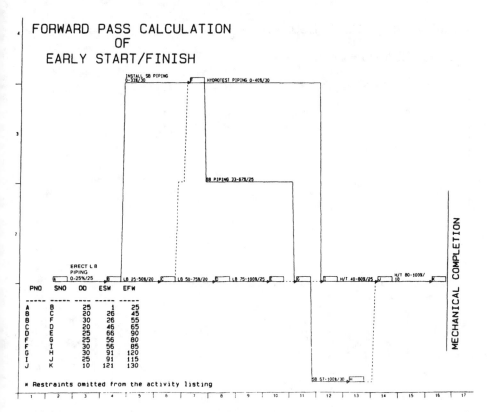

Figure 6.28 CPM example showing early start and early finish calculation.

still be monitored since significant slippage that occurs in excess of the current float would still impact the project completion date. Many project managers have been unpleasantly surprised by finding out too late that a "subcritical" task has become critical owing to lack of attention.

The other method of diagramming project networks is the *precedence diagramming method* (PDM), wherein activities are described within boxes rather than on arrows as shown previously in Fig. 6.30. Activity A1 is "install equipment foundations," with a duration of 65 days. When this activity is 50% complete, activity B1 can be *started*, but B1 cannot be *finished* until A1 is completed. (Activity B1 requires 40 days to complete.) When activity B1 is 50% complete, activity C1 can be started.

In practice, regardless of which network diagram method is used, data for CPM schedules are normally input to commercially available computer software packages for the calculation of the critical paths. These software packages allow for different duration units, different calendars, and variable holidays; they also permit the use of coding structures that serve as a focal point for other project scheduling areas of concern, such as resources and cost control.

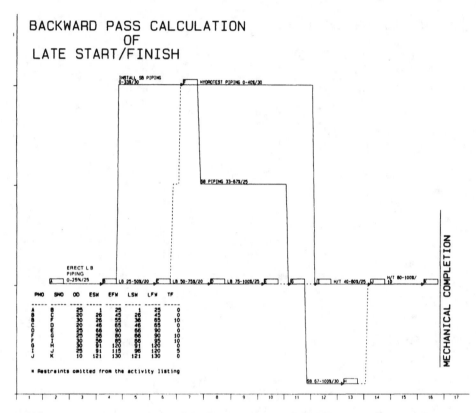

Figure 6.29   CPM example showing backward-pass calculation for latest start and latest finish.

When establishing activity durations, the project team members make some assumptions about resources based on their experience. To be sure that the project can be adequately supported, it is best to look at resources from three points of view. First, with dates constrained, when and in what quantities are resources required? Second, how can peak labor requirements be leveled out for better control of the project? And third, with a limited availability of resources, what completion dates are possible and/or reasonable? Although these studies can often be done on a manual basis for simple projects, it is not practical to do an extremely complex project without the aid of computers. Even for moderate projects, computer systems have made the process much easier. Using the developed network activities, the budgeted resources can be appropriately allocated by the project team, and anticipated resource require-ments over time can be readily calculated and displayed (as shown for labor in Fig. 6.11). The use of computerized project control systems also facilitates multi-project resource assessment in cases where a limited pool of resources must be shared between a number of different projects.

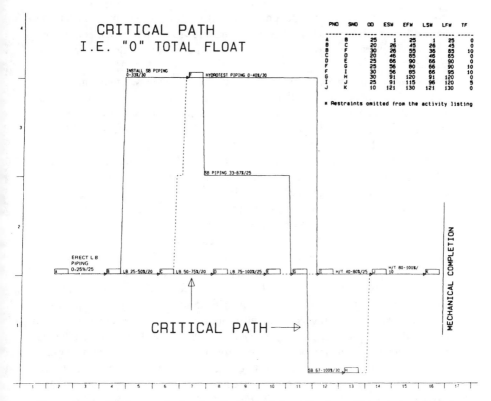

Figure 6.30 CPM example showing final calculation of early and late start, early and late finish, and available float.

With a scope in hand, with the organization necessary to manage that scope, with a plan to complete the scope, and with the necessary resources to support the plan, the project may be executed and its progress may be monitored. With large projects, diverse organizations, involved schedules, and complex monitoring, reports must be tailored to provide enough information so a user can guide and review project progress and yet avoid being overwhelmed by unneeded detail, as so frequently happens with computerized systems. The project planners must take an active role in reviewing tabulated reports and graphics with the end-users, and they must play a role in limiting and/or expanding the available data as necessary. With the coding, sorting, and selecting ability of most computerized CPM systems, it is easy to customize reports.

The simple Gantt chart is still a very suitable means of portraying project progress vs. time, and it can be readily generated by most available CPM software programs at any level of detail desired. Figure 6.8 is a detailed bar chart of an outage project in units of hours. Although bar charts are good overview aids, many supervisors prefer to have a list of tasks along with the schedule.

This list can be obtained for the entire project by using direct output from the CPM software, or it may be obtained by listing tasks summed up for an individual activity within the network, which also may be obtained directly from the software.

These lower level detailed schedules, or "punchlists," may in turn be individually computerized. In the case of material deliveries for large projects, the use of a spreadsheet or database in conjunction with a CPM model may be advisable. This would help ensure that procurement documents include all of the required equipment and also that the specified delivery dates are in concert with the construction installation schedule. Cross-referencing material and equipment to procurement documents, to construction areas, and/or to design documentation can be very helpful, even essential, during project implementation on large or fast-paced projects. This can be an immense aid to planning and managing when engineering, procurement, and construction all overlap and must be tightly coordinated with each other.

Progress reporting must also be tailored to the project and to the project team. Reporting can range from a manual table of progress by group over a 2-week period for a small project, as seen on Fig. 6.17, to a time-scaled chart of scheduled vs. target progress for each group within a large plant modernization project, as shown in Fig. 6.18. Graphical portrayals are excellent tools for quickly detecting problems. If problems are apparent, a more detailed analysis of the data may be required and more complex tabular reports may be generated, such as those from two refinery projects shown in Fig. 6.19 and 6.20. Using the coding capability of CPM software, a family of curves might be generated to depict different levels of data and different families of data for analysis. Figure 6.21 shows the progress for a total project, while Fig. 6.22 shows progress of four different aspects of the electrical contractor's effort; Fig. 6.23 shows indirect costs.

While many progress reports compare actual progress to planned progress, a more complete analysis is possible by analyzing costs or reviewing resource usage data. This allows a project manager to look at not just progress status but inefficiencies in the project as well, using earned value comparisons. This technique, which was discussed previously in Chapter 5, is commonly used in United States government projects under the Department of Energy's "Cost and Schedule Control System Criteria." It is commonly referred to as C/SCSC. With C/SCSC, comparisons are made between:

- Budgeted cost of work performed (BCWP), or earned value for work accomplished
- Budgeted cost of work scheduled (BCWS), work scheduled to be accomplished
- Actual cost of work performed (ACWP), costs incurred on the work accomplished

By looking at these additional factors, comparisons can be made to show whether resources are being applied as planned (ACWP vs. BCWS), and whether those resources are being used as efficiently as expected (ACWP vs. BCWP).

All reports, graphics, and available data are merely tools for the project team. Project control engineers must be an integral part of the team, and they must take an active role in reviewing the tools that are used to assess project variances and problems. They must participate in the formulation of "workarounds" with the team to bring about successful resolutions of conflicts and problems. Ultimately, they have the greatest impact upon a successful project completion.

RECOMMENDED READING

Bent, J. A. (1996). Humphreys, K. K., ed. *Effective Project Management Through Applied Cost and Schedule Control*. New York: Marcel Dekker.

CII (1990). *The Impact of Changes on Construction Cost and Schedule*. Publication 6–10, Austin, TX: Construction Industry Institute.

Gehrig, G. B., et al. (1990). *Concepts and Methods of Schedule Compression, Source Document 55*. Austin, TX: Construction Industry Institute.

Gido, J. (1985). *An Introduction to Project Planning*. 2nd ed. New York: Industrial Press, Inc.

Kerzner, H. (2002). Project management: a systems approach to planning. *Scheduling and Controlling*. 8th ed. New York: John Wiley & Sons, Inc.

Kimmons, R. L. (1990). *Project Management Basics*, New York: Marcel Dekker, Inc.

7

Cost Indexes, Escalation, and Location Factors

7.1 COST INDEXES

A typical problem faced by an engineer is estimating the current or future cost of equipment, plants, or buildings. One way to make such estimates is to obtain costs of similar projects from an earlier date and update these costs to the present time.

Prices of equipment, plants, or buildings vary from time to time in accordance with competitive market conditions and the general state of inflation or deflation of a country's currency. They also vary at any given time from one area of the country to another area and from one country to another country. These differences in price from one time to another time period and from one location to another can be measured by cost indexes.

A *cost index* is the ratio of cost or price for a given commodity or service or set of commodities or services at a given time and place compared to the cost or price at a base or standard time and place. There are many cost indexes published by many organizations and magazines, and they cover practically every area of interest to an engineer. There are indexes for building construction, for wage rates for various industries and types of plants, and for various types of equipment, material, or commodities.

Remember, cost indexes are based on present costs compared with cost history. As published, they usually do not forecast future escalation; they leave that up to the discretion of the individual user.

Table 7.1 gives examples of some published cost indexes. These indexes have different base years. For example, the *Engineering News-Record* construction cost index has a base year of 1913 = 100 and the Marshall and Swift equipment cost index has a base year of 1926 = 100. The choice of the base year for a cost index is not important because the current values of any index simply

Table 7.1   Selected Cost Indexes[a]

Year	Engineering News-Record construction cost index (Base 1913 = 100)	Chemical Engineering plant cost index (Base 1957/ 59 = 100)	Marshall & Swift Industrial equipment cost index (Base 1926 = 100)	Consumer price index[b] (Base 1967 = 100)
1982	3825	314	746	289
1983	4066	317	761	298
1984	4146	323	780	311
1985	4195	325	790	322
1986	4295	318	798	328
1987	4406	324	814	340
1988	4519	343	852	354
1989	4615	355	895	371
1990	4732	358	915	391
1991	4835	361	931	408
1992	4985	358	943	420
1993	5210	359	964	433
1994	5408	368	993	444
1995	5471	381	1028	457
1996	5620	382	1039	470
1997	5826	387	1057	481
1998	5920	390	1062	488
1999	6059	391	1068	499
2000	6221	394	1089	516
2001	6343	394	1094	530
2002	6538	396	1104	536

[a]Index values are rounded to nearest integer value.
[b]The weightings within the Consumer Price Index were changed in mid-1998. Values shown for earlier years have been adjusted to the new weightings.

represent current cost as compared to cost in the base year. For example, if an index has a current value of 425, the current cost of items reflected in that index is 425% of the base year cost.

For purposes of comparison, the indexes in Table 7.1 were converted to 1990 as base 100 in Table 7.2. To convert an index to a new base year, divide the index to be converted by the index value of the new base period (on the old or previous base) and multiply by 100. For example, to convert the *Engineering News-Record* construction index to a new base of 1990 = 100, divide all the index values by 4732 (year 1990). Similarly, to convert the *Chemical Engineering* plant cost index to 1990 = 100, divide by 358 (year 1990). When cost indexes are converted to the same base years and are plotted together for comparison,

Table 7.2 Selected Cost Indexes Corrected to 1990 Base Year[a]

Year	*Engineering News-Record construction cost index*	*Chemical Engineering Plant Cost index*	Marshall & Swift installed equipment cost index	Consumer price index
1982	81	88	82	74
1983	86	89	83	76
1984	88	90	85	80
1985	89	91	86	82
1986	91	89	87	84
1987	93	91	89	87
1988	95	96	93	91
1989	98	99	98	95
1990	100	100	100	100
1991	102	101	102	104
1992	105	100	103	107
1993	110	100	105	111
1994	114	103	109	114
1995	116	106	112	117
1996	119	107	114	120
1997	123	108	116	123
1998	125	109	116	125
1999	128	109	117	128
2000	131	110	119	132
2001	134	110	120	136
2002	138	111	121	137

[a]Rounded to nearest integer value.

the curves vary irregularly and may not correlate. Costs of various items simply do not change over time in the same manner. The results from Table 7.2 are shown in Figure 7.1, and it can be seen that the correlation is poor.

7.1.1 How to Use an Index

Cost indexes are based on ratios and therefore should not be subtracted. To update the cost from year A to year B, multiply year A's cost by the ratio of year B's index over year A's index. For example, using the Marshall and Swift equipment cost index, an index of average installed costs of process equipment in 47 industries, say that equipment installed in 1999 cost $15,000,000. Approximately what would the same equipment have cost if installed in 2002?

$$\$15,000,000 \times \frac{1104}{1068} \approx \$15,501,000 \text{ (2002 cost)}$$

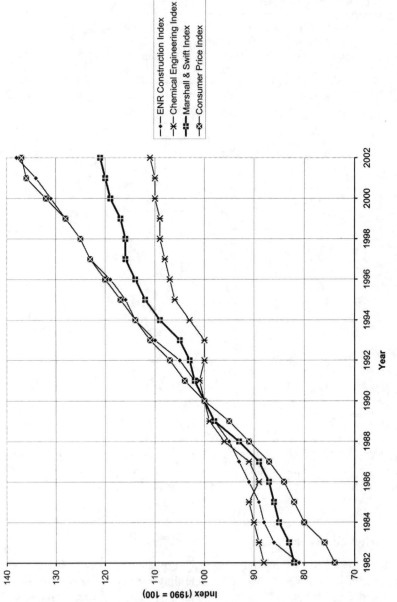

Figure 7.1 Comparison of selected cost indexes.

The use of cost indexes has several limitations, and caution is advised in their use. To cite these potential limitations, one must bear in mind the following:

1. Cost indexes are limited in accuracy. Two indexes could produce two different answers.
2. Cost indexes are based on average values. Specific cases may be different from the average.
3. At best, accuracy may be limited by applying indexes over a 4- to 5-year period.
4. For time intervals over 5 years indexes are highly inaccurate and should be used for order-of-magnitude estimates only.
5. Due to technology changes or other factors, the method by which any given index is complied may periodically be changed by the complier. In such cases, index values before and after the date of the change may not be directly comparable.

Judgment also has to be used as to whether the index applies to the specific cost being updated. If the specification or scope for the new cost has changed, the index may not apply.

7.1.2 How an Index is Compiled

The published indexes are compiled in various ways. Although they differ in nature, they have one thing in common. They are compiled for a specific set of conditions or a specific industry, type of equipment, building, or plant. They can be simple or complex. The *Engineering News-Record* construction cost index is based on the prices of certain fixed quantities of cement, lumber, structural steel, and common labor. The Marshall & Swift equipment cost index averages 47 industries. Marshall & Swift also provides specialized industry indexes, such as cement, chemical, glass, paper, petroleum, clay products, rubber, paint, electrical power equipment, mining, refrigeration, and steam power.

The *Chemical Engineering* chemical plant cost index is a good example of a complex index. It is made up of 41 Bureau of Labor Statistics (BLS) Producer price Indexes, 12 BLS labor cost indexes, and a calculated labor productivity factor. It consists of four major components weighted as follows:

Equipment	50.675%
Construction labor	29.000%
Buildings	4.575%
Engineering and supervision	15.750%
	Total = 100%

The four components are weighted in the index based upon information obtained in a survey of CPI companies, engineering firms, index publishers, and technical organizations.

The equipment category is made up of seven subcomponents:

Heat exchangers and tanks	33.8%
Process machinery	12.8%
Pipes, valves, and fittings	19.0%
Process instruments	10.5%
Pumps and compressors	6.4%
Electrical equipment	7.0%
Structural supports and miscellaneous	10.5%
	Total = 100%

The *Chemical Engineering* index was created in 1963, was revised in 1982, and was updated in 2002 to incorporate technological and construction labor-productivity changes since the 1982 revision. The 1982 version of the *Chemical Engineering* index incorporated 66 BLS price indexes, some of which have been discontinued. This also contributed to the need for the change in the *Chemical Engineering* index.

To show how an index is compiled and updated, let us evaluate a hypothetical index constructed for a chemical plant. For calculation of the total plant cost index, assume that the components are divided into the following weighted categories:

Fabricated equipment—columns, vessels, tanks, heat exchangers, and fired heaters.

Process machinery—pumps, compressors, blowers, mixers, and instruments.

Bulk materials—pipe, valves, fittings, structural steel, concrete, insulation, and electrical components.

Construction labor—the craftpersons working in the field.

Engineering—the design, procurement, expediting, estimating, and cost control people working predominantly at the contractor's home office, plus the field supervision staff and the contractor's indirect field costs.

The percentages listed below are based on the dollar mix for, say, 2000. The hypothetical index is to be calculated for February 2003 based on the relative growth of the individual mixes for each category. The February 2003 plant index calculates as follows:

	2000 mix × Escalation factor	Feb. 2003 index
Fabricated equipment	22 × 328 =	72
Process machinery	10 × 321 =	32
Bulk materials	24 × 371 =	89
Construction labor	28 × 270 =	76
Engineering	16 × 316 =	51
	100%	320

If the 2000 index was 261, the plant cost index for February 2003 indicates an increase of 22.6% (320 divided by 261).

The individual indexes for each category are derived by applying escalation factors to the 2000 values. These factors are obtained either through analysis of prices recently paid by the company's purchasing department or through published component indexes, such as those published by the Bureau of Labor Statistics.

7.1.3 Analysis of some Building Cost Indexes

Shellenberger (1967) gave an excellent discussion of building cost indexes. Despite the passage of time, the paper is basically valid today. The following discussion paraphrases portions of that paper entitled "A Comparison of Building Cost Indexes."

One of the important functions of cost indexes is to afford the estimator a means of upgrading the cost of a similar facility from the past to the present. Provided that the estimator uses discretion in choosing the proper index, a reasonable approximation of the present cost should result.

Only a limited amount of information has been published about many of the cost indexes that are available. A very brief description and data for some of them are published in the "Quarterly Cost Roundup" issues of *Engineering News-Record*. More detailed information is available in the AACE International *Cost Engineers' Notebook* which is periodically updated. Also, any available description of the index being used should be checked, a seemingly obvious precaution that is often overlooked during a rush job. It is the purpose of this discussion to present additional information about the building cost indexes and briefly to compare them and mention points to be considered when using them.

A large number of cost indexes are available to the estimator; in building construction alone, at least 15 major indexes are compiled in the United States covering from 1 to 17 types of buildings in from 1 to over 300 different locations. Specialized construction indexes are also available covering railroad stations, airplane hangers, utility buildings, and other types of structures.

The well-known *Engineering News-Record* construction cost and building cost indexes are frequently used erroneously. Many estimators fail to realize that the *ENR* building cost index was *not* designed as a cost index for the construction of buildings. The two *ENR* indexes are essentially identical in that they are based on building a hypothetical block of construction in 20 United States cities and 2 Canadian cities. The indexes differ only in that the building cost index incorporates skilled labor wages and fringe benefits, and the construction cost index uses common labor.

The differences between indexes are the combined results of many things, but mainly they reflect differences in the mix of materials, labor, engineering costs, and so on. To illustrate these differences, major components included in three building cost indexes for specific areas are presented in Table 7.3. Of the 20 separate items

Table 7.3   Items Included in Building Cost Indexes

Index	Materials				Labor						Land			Arch. and Eng.						
	Cost	Sales Tax	Freight	Expediting	Cost	Efficiency	Overtime	Premiums	Procurement	FICA and comp.	Cost	Excav. and back.	Yard improv.	Cost	Fee	Constr. exp.[a]	Contractors fee	Bidding competition	Future Price trends	New concepts[b]
Aberthaw (I Type — New England)	X	–	X	X	X	X	–	–	–	X	–	X	X	–	–	X	–	X	–	X
Austin (I Type — Central and Eastern)	X	X	X	X	X	X	–	–	X	X	–	X	–	X	X	X	X	–	–	X
Fruin – Colnon (5 Types – St. Louis)	X	X	X	X	X	X	–	–	X	X	–	X	X	–	X	X	X	X	–	

Group 1 – industrial buildings in specific areas.

[a]Includes field supervision, field office, temporary construction, equipment rental, tools, insurance, etc.

[b]Modifications to account for new design concepts, new materials, or new construction methods.

shown, only 8 are taken into consideration by all three indexes; two important labor cost factors are not considered by any of the three. Table 7.4 presents a similar comparison of the cost of general buildings in specific areas for three indexes. Seven of the 20 items are taken into account by all three indexes, and 14 are considered in developing the Turner and the Smith, Hinchman, and Grylls indexes. Table 7.5 is a partial comparison of a third group of indexes which use national average

Table 7.4   Items Included in Building Cost Indexes

Index	Materials				Labor						Land			Arch. and Eng.						
	Cost	Sales Tax	Freight	Expediting	Cost	Efficiency	Overtime	Premiums	Procurement	FICA and comp.	Cost	Excav. and back.	Yard improv.	Cost	Fee	Constr. exp.[a]	Contractors fee	Bidding competition	Future Price trends	New concepts[b]
Fuller (Eastern Cities)	X	X	X	–	X	X	–	–	–	X	–	X	–	–	–	X	–	–	–	X
Smith, Hinchman and Grylls (Detroit, Mich.)	X	X	X	X	X	X	X	X	X	X	–	X	X	–	–	X	X	X	X	X
Turner[c] (Eastern Cities)	X	X	X	X	X	X	–	X	–	X	–	X	X	–	–	X	X	X	X	–

Group 2 – General buildings in specific areas.

[a]Includes field supervision, field office, temporary construction, equipment rental, tools, insurance, etc.

[b]Modifications to account for new design concepts, new materials or new materials or new construction methods.

[c]The efficiency of plant and management are also considered in this index.

Table 7.5 Items Included in Building Cost Indexes

Index	Materials — Cost	Sales tax	Freight	Expediting	Labor — Cost	Efficiency	Overtime	Premiums	Procurement	FICA and comp.	Land — Cost	Escov. and back.	Yard improv.	Cost	Arch. and Eng. — Constr. exp.[a]	Contractors fee	Bidding competition	Future price trends	New concepts[b]
American Appraisal (4 Types — 30 Cities)	X[c]	—	X	—	X[c]	X	—	—	—	—	—	—	—	—	X	X	—	—	—
Boeckh (10 Types — 57 Areas)	X	X	—	—	X	X	X	X	X	X	—	—	—	X	X	X	X	—	—
Campbell (5 Types — 17 Areas)	X[c]	—	—	—	X[c]	—	—	X	—	—	—	—	—	—	—	—	—	—	X
Dow (F. W. Dodge) (17 Types — 237 Areas)	X[d]	—	X	—	X[d]	—	—	—	—	X	—	—	—	—	—	—	X	—	—
M. & S. (4 Types — Many Areas)	X	—	—	—	X	—	—	—	—	—	—	—	—	—	—	—	—	—	X

Group 3 – Various types of buildings in many areas.

[a] Includes field supervision, field office, temporary construction, equipment rental, tools, insurance, etc.

[b] Modifications to account for new design concepts, new materials or new materials or new construction methods.

[c] The efficiency of plant and management are also considered in this index.

figures. Components included in each of five indexes are presented in the table. Of the 20 items listed, only two are considered by all five companies, and even in these items there are major differences, as indicated by the footnotes.

From Tables 7.3, 7.4, and 7.5, it is obvious that the indexes vary, in some cases quite widely. In selecting a given index to upgrade an estimate, one should first consider the type of building, the area, and then the individual components included in the original estimate. If some major item was not included, judgment must be made as to how much it changed compared to the index, and if a major difference is apparent, a correction should be made for this. The estimator should not fail to appreciate the significant effect the original estimate date may have on the final number.

7.1.4 Limitations in Using Published Indexes

There is no perfect index. Although a large amount of time may go into publishing an index, any index has limitations.

1. A cost index is a statistically weighted composite average and suffers from all the disadvantages of averages. Although the average cost per cubic yard for installing concrete may be known, this does not mean that any one foundation will cost exactly that amount per cubic yard.

2. Most indexes represent reproduction costs and do not account for any radical technological changes in design or construction. A weighting based on the mix of materials 20 years ago will hardly be representative of today's construction, not just in terms of material and labor ratios, but also in terms of the materials themselves. Therefore any fixed base weight index is doomed to obsolescence in today's rapidly changing technology. Periodic restudy of the index components and mix is required.

3. A reporting time lag exists because it takes several months to collect, analyze, and publish the information. This delay is especially prevalent when an index is based in part on indexes published by other organizations. For example, the Bureau of Labor Statistics is undoubtedly the primary source of information in the United States and is used in many other indexes; yet the BLS index runs 2 or more months late. In summary, most indexes are about 3 to 6 months behind time. An exception to this is the *Engineering News-Record* construction index. Because of its simplicity, pricing of the 3 basic commodities plus labor is on a relatively current basis.

4. Indexes lack sensitivity to short-term economic cycle swings. Most indexes are based on published list prices rather than actual market prices, and average labor conditions rather than actual conditions. The realistic index will appear somewhat like a roller coaster superimposed on a long-term creeping inflation trend.

Several significant variables contribute to abnormal increases. In a period of rising construction employment, labor productivity drops as marginal workers enter. When the bottom of the barrel is reached, overtime is used, with its premium labor costs. Then with labor in tight supply, wage rate contract negotiations become more difficult, and abnormally high wage concessions may result. The same is true for materials. As the economy is booming, material costs escalate more rapidly as the economy moves into a seller's market. List prices may not move up, but more importantly, discounts to aggressive purchasing departments disappear; and premium costs are paid to ensure deliveries within reasonable time schedules. As the boom peaks and the trend reverses, costs of construction actually drop, even though the drop may not show up in indexes. The base labor wage rates do not drop, but as overtime disappears and labor productivity returns to average or better than average, absolute labor costs decrease. Equally the return to tight competitive bidding conditions by material suppliers brings about cost reductions in materials. If the market gets tight enough, vendors bid either work or equipment at out-of-pocket costs with very little or no profit margin. Then the cycle starts all over again.

As stated earlier, most published indexes do not forecast future escalation. The *Engineering News-Record* indexes do forecast the escalation for the following year, but not beyond one year.

A simple way to project escalation is to plot historical cost index trends and then extrapolate them into the future. This requires a considerable degree of judgment when forecasting escalation for construction projects. Some companies try to gain insight into the movements of a construction cost index by comparing its history to broad price indexes for the entire economy. Some broad price indexes are:

1. Gross national product deflator
2. Plant and equipment deflator
3. Producer price index
4. Consumer price index

These indexes are maintained by the United States Bureau of Labor Statistics.

Periodically government economists predict escalation rates for these broad price indexes. These predictions may in some cases be applicable to construction cost indexes.

7.1.5 Procedure for Estimating Escalation

It is not advisable to apply escalation to the overall index. A good approach is to analyze the movements of the components of the index. Then, by applying escalation rates to each component, calculate an overall composite escalation.

Following is a suggested procedure:

1. Determine the percentage cost breakdown for the components of the index.
2. Determine the rate of escalation for each component. This rate can be based either on judgment or published forecasts.
3. Determine the schedule for the construction project. This step is perhaps the key to this approach. Step 3 is perhaps the key to this approach. The calculation is meaningless unless a reasonable judgment is made of the time interval applicable to each component because components do not start and finish at the same time. After the design is well underway, the material components are purchased. Construction labor and subcontractors start after the material is delivered to the job site. A similar sequence follows for the finish of the components.

Example

A possible breakdown for a construction project is (1) materials, (2) construction labor, (3) subcontracts, and (4) engineering costs, including overhead and profit. Assume the following parameters:

1. Estimated cost is $25 million–based on today's costs.
2. Breakdown of the components: material, 50%; construction labor, 25%; subcontracts, 10%; and engineering costs, 15%.
3. Schedule per Figure 7.2. The time period for each component starts from a common point: the time of estimate or today's cost, which is referred to in Figure 7.2 as base time. All times for escalation

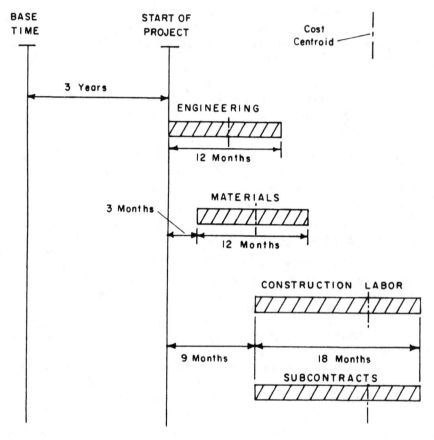

Figure 7.2   Project time Schedule.

calculation are taken to the centroid of the component (ie, time when half of the costs have been spent). In this example, for engineering and material costs, the cost centroid is about halfway through the time period; for construction labor and subcontracts, the centroid is approximately two-thirds of the way through the time period. Relative to base time, the cost centroids are as follows:

   a. Engineering: 3 years + ($\frac{1}{2}$) (12 months) = 3.5 years

   b. Material: 3 years + 3 months + ($\frac{1}{2}$) (12 months) = 3.75 years

   c. Construction labor: 3 years + 9 months + (2/3)(18 months) = 4.75 years

   d. Subcontracts: same calculation as labor = 4.75 years

  4. Rate of escalation: materials, 6% per year; construction labor, 10% per year; contracts, 8% per year; and engineering, 5% per year.

Therefore the total project cost, including escalation, is:

1. Engineering: $\$25 \times 10^6 \times 15\% \times (1.05)^{3.5} = \$4.45 \times 10^6$
2. Material: $\$25 \times 10^6 \times 50\% \times (1.06)^{3.75} = 15.55 \times 10^6$
3. Construction labor: $\$25 \times 10^6 \times 25\% \times (1.10)^{4.75} = 9.83 \times 10^6$
4. Subcontracts: $\$25 \times 10^6 \times 10\% \times (1.08)^{4.75} = \underline{3.60 \times 10^6}$
5. Total estimated cost $= \$33.43 \times 10^6$

The escalation added to the project is

$$\$33.43 \times 10^6 - \$25 \times 10^6 = \$8.43 \times 10^6$$

Note that the rates have been compounded in calculating the escalation. The effect of compounding is relatively insignificant for short periods, and errors in the assumed escalation rates and breakdown probably introduce greater differences. For longer periods, compounding can be significant, especially if several years are involved. The accepted practice is to calculate escalation on a compounded basis.

7.2 LOCATION FACTORS

Perhaps the most common technical inquiry received by AACE International, and also by the International Cost Engineering Council, goes something like this: "I am trying to evaluate the costs of constructing a new plant/facility/building in three foreign countries, and I'm having trouble locating relevant international cost data. Do you have any suggestions?" This is quickly followed by ..."My feasibility report is due tomorrow," and, "I know what it costs to construct this facility in my country, but I need to be able to adjust that cost to these other countries. Can you help me?"

While cost engineers generally are very familiar with major sources of cost data in their own country, it is not surprising that they often are unaware of useful sources of commonly published cost data and related information in other countries. This problem can be further compounded by lack of time to perform a proper search, publications being in various languages, and lack of information about key factors that can impact the estimate for particular geographic locations.

In this regard, a word of caution is necessary. Published information must always be used with care! Every location factor or commonly available cost index has its own underlying method of construction, with its particular inherent components and weightings. It is vital for the estimator using such quick-estimate data to understand how they were created and to recognize just what their limitations and applications are. In addition, published data also are often inadequately explained and frequently improperly dated. Date of publication is meaningless because the data may be months or years old and may require adjustment to current cost levels. Equipment cost data may or may not include ancillaries and/or transportation and installation, etc. Too often it seems that in the rush to complete the assignment, people will grasp any number they can find without fully understanding how it was derived, or what it represents.

With location factors one must recognize that they generally reflect only the relative cost to replicate a facility *exactly* in another location. The factors do not consider cost effects which are introduced by site-unique conditions such as climate, earthquake and geological considerations, etc. If the design is not identical in both locations, the cost differences are not generally accounted for if location factors alone are used.

7.2.1 Background Sources

In order to make a proper estimate of a particular international project, it is vital that the cost engineer understand the conditions existing in the country or countries where a project is to be located. In a paper presented at the Fifth International Cost Engineering Congress, Utrecht, Netherlands, Walker (1978) outlined the major economic system parameters to be evaluated as follows:

- Political
  Stability
  Attitude towards foreign investment
  Type of bureaucracy
- Finance
  Banking system
  Insurance regulations
  Tax system
  Duties
- Legal system
  Laws governing conduct of business and individual freedom
- Social system
  Business ethics
  Education
  Language and religion
- Geography
  Infrastructure and communication
  Climate
- Industry
  Capacity
  Diversity
  Efficiency

Walker expressed the interactions of these factors in matrix form as shown in Figure 7.3.

Many others (see recommended reading and other information sources at the end of this chapter) have discussed site- or country-specific factors that can

Area of Influence	International	National	Local	Project	Company
Economic System					
Political	x	x	x		x
Financial	x	x		x	x
Legal		x	x	x	x
Social	x	x	x		x
Geographic		x	x	x	
Industrial	x	x	x		

Figure 7.3   Economic basis of cost: the Levels at which the economic systems impact on cost. (From Walker, C. G. Estimating Construction Costs Abroad, *Transactions Fifth International Cost Engineering Congress*, Dutch Association of Cost Engineers, The Hague, pp. 40–46, 1978.)

impact the cost, schedule, and/or price for an international project. These include: local material quality and availability, labor availability, equipment availability, labor productivity, import duties, import licenses, customs, local taxes, language, length of workweek, holidays, inflation, fluctuating exchange rates, religious customs, buy-local laws, shipping cycles, weather and climatic impacts, workforce level of education, communications, language, logistics, workforce housing, training, and many other relevant factors. Regional variations of these factors within a country must also be expected, and remoteness and distance from major cities or supply centers can often aggravate the above problems even further.

Patrascu (1979, 1988) has proposed preestimate survey checklists to help identify background concerns for foreign construction projects. These checklists delineate a large number of factors which must be considered, including those described by Walker (1978). A detailed discussion of this topic and a checklist of factors which should be considered before undertaking any international or multinational project is provided in Chapter 10.

For offshore projects an excellent detailed checklist (ACostE, 1982) is available from the Association of Cost Engineers (ACostE), Lea House, 5 Middlewich Road, Sandbach, Cheshire, CW11 9XL, United Kingdom. Based on North Sea oil field construction experience, this publication provides a very detailed breakdown for offshore work.

Many other references are useful in preparation for estimating international projects. However background literature cannot replace pre-estimate site visits, proper contract development, and conversations with others who have experience estimating work in the particular country or countries of interest. Review of appropriate literature, however, can help to ensure that all important factors for the project have been considered in developing the estimate.

The topics mentioned and those listed in Chapter 10, while far from forming a complete list, should nevertheless make the cost engineer aware of most of the potentially important considerations which are unique to international work.

7.2.2 International Location Factors

When little time is available or warranted to perform the type of background studies suggested above, and detailed design and engineering has not been completed, estimators must turn to published indexes, location factors, or other sources of relevant data for help. Table 7.6 shows a collection of sources for multicountry information. The references in Table 7.6 are examples of multicountry data sources. For conceptual studies these may be useful, depending on what countries are of interest. Various reports by banks, governments, trade associations, etc also exist if one is willing to search for them. One example is a report (Fong, undated) that compares building costs in a number of countries to those in Malaysia. This special report was issued by a government task force and provides a handy reference for those comparing building costs in Pacific rim countries.

Massa (1984, 1985) has proposed development of international cost location factors based upon a weighting of 33.05% for a labor factor, 53.45% for an equipment and civil material factor, and 13.50% for an indirect and home office cost factor. He presents a detailed form for calculating these three factors and the composite factor for any given country referenced to US Gulf Coast costs. Massa has also provided labor factors for many countries. His factors are presented in Table 7.7 along with a list of labor factors complied by Bent (1996) and country location factors previously reported by Bridgewater (1979). The Bridgewater factors are for complete chemical plants and are referenced to both the United Kingdom and the United States. Note that the Bridgewater factors reflect currency exchange rates, taxes, and duties which were prevailing when the factors were first reported. They should not be used without adjustment for changes which have occurred since the factors were developed.

Also included in Table 7.7 are selected 2003 location factors for some major cities from Aspen Richardson's *International Construction Cost Factor Location Manual* which is published annually on CD-ROM (see Appendix C). See the source for details on compilation of these factors and the cities to which they apply. This publication is regularly updated and as information becomes available additional countries are added.

In addition to these sources, another excellent source of location factor information is the AACE International Cost Estimating Committee. The committee maintains a listing of industry contacts who are willing to share location factors and foreign cost estimating data.

Table 7.6 International Data Sources–Multiple Country

Foreign Labor Trends—periodic reports of labor trends and costs for specific countries— includes key labor indicators, information on unionization, labor availability, recent developments affecting the workforce, etc. Each report covers one specific country and is prepared by the American Embassy staff in that particular country. (U. S. Department of Labor, Bureau of International Affairs, Washington, DC)

Dutch Association of Cost Engineers Prijzenboekje (Price Book)—periodically published booklet of actual installed project costs as reported by about 200 Dutch companies. Although the booklet is published in Dutch, an English language glossary of terms makes it easy to use but those who do not read the Dutch language. The data provided is valid for the European Economic Community. (Dutch Association of Cost Engineers, P.O. Box 1058, 3860 BB Nijkerk, the Netherlands)

Hanscomb/Means International Construction Cost Intelligence Report—newsletter— provides comparative building construction cost information for many countries. (Hanscomb, Faithful & Gould, 817 W. Peachtree St. NW, Suite 500, Atlanta, GA 30308)

U. N. Monthly Bulletin of Statistics—includes a variety of production, trade, financial, commodities, construction, wage and other cost/price indices and statistics for about 190 countries.(United Nations, New York, NY)

Engineering News Record—weekly magazine - primarily US and Canada, variety of commercial/industrial construction and builders indexes plus materials prices and labor rates; substantial North American data featured in "Quarterly Cost Roundup" issues; World Parameter Costs column features building costs for one or two countries per quarterly issue. (McGraw-Hill, Inc., 330 W. 42nd Street, New York, NY 10036)

Costos de Construction Pesada y Edificacion (Heavy Construction and Building Costs)— cost estimating data base on building, industrial and heavy construction reflecting costing in Mexico and other Latin American countries eg, Bolivia, Venezuela, Panama, Brazil and Chile. (Bimsa Southam, S.A. de C.V., Sófocles No. 118, Col. Chapultepec Polanco, Deleg. Miguel Hidalgo, C.P. 11560 Mexico DF, Mexico)

Spon's Architects' and Builders' Price Book—contains a European section for tendering and costs of labor and materials in 13 countries. *Spon's European Construction Costs Handbook* provides coverage of 28 countries in Europe plus the United States and Japan. *Spon's Asia Pacific Construction Costs Handbook* provides similar coverage of 15 countries in Asia plus the United States and the United Kingdom. *Spon's Middle East Construction Price Book* is a two volume set which provides detailed unit cost information for 6 Middle East countries. (E. & F. N. Spon, Ltd., Spon Press, 11 New Fetter Lane, London EC4P 4EE, United Kingdom)

R. S. Means Co.—various cost books and CD-ROM databases published annually for building and industrial construction in the U.S. and Canada. (R. S. Means Co., 100 Construction Plaza, Kingston, MA 02364)

Aspen Technology Inc.—*Aspen Richardson's R-Books* unit cost databases published for U.S. and Canada building and general construction; U. S. And Canada process plant construction; and international construction. (Aspen Technology, Inc., Ten Canal Park, Cambridge, MA 02141-2201)

International Construction Costs and Reference Data Yearbook—provides detailed information on construction costs, including fully loaded labor rates, for 30 countries. (John Wiley & Sons, New York, 1996)

Table 7.7 Sample Location Factors

Location	Massa's Labor Productivity Factors US = 1.0	Bridgewater's Factors for Chemical Plants[a] UK = 1.0	Bridgewater's Factors for Chemical Plants[a] US = 1.0	Bent's Labor Productivity Factors US = 1.0[c]	Aspen Richardson 2003 Cost Factors UK = 1.0
Algeria	1.82	—	—	—	—
Argentina	2.00 (1.30–2.60)	—	—	—	1.10[b]
Australia	1.20 (0.96–1.45)	1.4	1.3	1.6	1.15–1.22
Austria	1.60 (1.57–2.10)	1.1	1.0	—	—
Belgium	1.14 (1.14–1.50)	1.1	1.0	1.3	1.16
Brazil	—	—	—	1.8	1.12[b]
Canada	—	1.25	1.15	1.2	—
East	1.14 (1.08–1.17)	—	—	—	0.97–1.00
West	1.07 (1.02–1.11)	—	—	—	0.97–0.99
Newfoundland	—	1.3	1.2	—	—
Central Africa	—	2.0	2.0	—	—
Central America	—	1.1	1.0	—	—
Ceylon	3.50	—	—	—	—
Chile	2.70 (2.00–2.90)	—	—	—	—
China	—	—	—	2.2	1.05–1.14
imported element	—	1.2	1.1	—	—
indigenous element	—	0.6	0.55	—	—
Colombia	3.05	—	—	—	1.09
Denmark	1.28 (1.25–1.30)	1.1	1.0	—	—
Eastern Europe	—	—	—	2.0	—
Egypt	2.05	—	—	—	1.07
Finland	1.28 (1.24–1.28)	1.3	1.2	1.7	—
France	1.52 (0.80–1.54)	1.05	0.95	1.3	1.11
Germany, West	1.20 (1.00–1.33)	1.1	1.0	1.1	1.09

Ghana	3.50	—	—	—	—
Greece	1.49	1.0	0.9	—	—
Guatemala	—	—	—	—	—
Hong Kong	—	—	—	1.5	1.10
India	4.00 (2.50–10.0)	—	—	2.5	1.00
imported element	—	2.0	1.8	—	—
indigenous element	—	0.7	0.65	—	—
Indonesia	—	—	—	1.9	1.11[2]
Iran	4.00	—	—	—	—
Iraq	3.50	—	—	—	—
Ireland	—	0.9	0.8	1.65	1.28
Israel	—	—	—	1.8	—
Italy	1.48 (1.10–1.48)	1.0	0.9	1.4	1.12
Japan	1.54 (1.00–2.00)	1.0	0.9	1.1	1.23
Kuwait	—	—	—	2.1	1.07
Malaysia	—	0.9	0.8	1.9	1.11
Mexico	1.56 (1.54–3.15)	—	—	1.5–1.8	1.01
Middle East	—	1.2	1.1	—	—
Morocco	—	—	—	—	1.12
Netherlands	1.25 (1.25–1.60)	1.1	1.0	1.35	1.17
New Zealand	—	1.4	1.3	1.5	—
Nicaragua	2.67	—	—	—	—
Nigeria	2.22	—	—	—	1.12
North Africa	—	—	—	—	—
imported element	—	1.2	1.1	—	—
indigenous element	—	0.8	0.75	—	—
Norway	1.23	1.2	1.1	1.75	—

(continued)

Table 7.7 *Continued*

Location	Massa's Labor Productivity Factors US = 1.0	Bridgewater's Factors for Chemical Plants[a] UK = 1.0	Bridgewater's Factors for Chemical Plants[a] US = 1.0	Bent's Labor Productivity Factors US = 1.0[c]	Aspen Richardson 2003 Cost Factors UK = 1.0
Pakistan	—	—	—	2.2	—
Peru	—	—	—	—	1.21
Philippines	2.86	—	—	2.5	1.16
Poland	—	—	—	1.9	0.93
Portugal	1.66	0.8	0.75	—	—
Puerto Rico	1.54	—	—	—	—
Russia	—	—	—	2.0	1.51[b]
Saudi Arabia	—	—	—	1.6	1.03
Singapore	4.00	—	—	1.6	1.08
South Africa	1.58	1.25	1.15	1.4–1.9	1.05
South America (north)	—	1.5	1.35	—	—
South America (south)	—	2.5	2.25	—	—
South Korea	—	—	—	1.3	1.11
Spain	1.74	—	—	1.7	1.03

imported element	—	1.3	1.2	—	—
indigenous element	—	0.8	0.75	—	—
Sri Lanka	—	—	—	2.5	—
Sweden	1.18 (1.10–1.20)	1.2	1.1	1.35	—
Switzerland	—	1.2	1.1	1.5	—
Taiwan	1.52 (1.52–7.20)	—	—	1.3	0.96
Thailand	2.82	1.1	—	—	1.25
Turkey	2.32	—	1.0	—	0.82
United Arab Emirates	—	1.1	—	1.7	1.06
United Kingdom	1.53 (0.70–2.46)	1.0	0.9	1.5	1.25
United States	1.00	1.1	1.0	1.2–1.6[d]	1.00
Venezuela	2.00	—	—	1.65	0.99[b]
Vietnam	—	—	—	—	1.15
Yugoslavia	1.40 (1.25–2.03)	1.0	0.9	—	—

[a]Increase chemical plant factor by 10% for each 1000 miles or part of 1000 miles that the new plant location is distant from a major manufacturing or import center or both. When materials or labor, or both, are obtained from more than a single source, prorate the appropriate factors. Factors do not consider investment incentives.

[b]Rate of exchange and inflation are very volatile. Use factors with care.

[c]Nonunion, southeast United States.

[d]Union locations.

Table 7.8 Country Indexes[a]

Country/Index	Comments
Australia	
Building Cost Index	Construction labor and material costs
Construction Cost Index	Weekly earnings; building and nonbuilding materials
Australian Builder	Price information on raw materials
Cordells Building Cost Book	Price information for non-residential building
Monthly Summary of Statistics	Manufacturing articles: materials; building/non-building materials; metallic materials; wage rates indices
Brazil	
Revista de Precos	Cost/price indices for residential and non-residential construction; Portuguese
Boletim de Custos	Cost/price indices for residential and non-residential construction; Portuguese
A Construcao	Residential and non-residential construction and project costs; Portuguese
Conjuntura	Indices for industrial machinery and equipment; residential and non-residential construction material; Portuguese
NTC—Associacao Nacional dos Transportadores de Cargas	Transportation rates for industrial materials; Portuguese
Canada	
Statistics Canada	Variety of cost/price indexes for construction costs and capital expenditures; English and French
England	
A.C.E. Indices of Erected Plant Costs	Indices of erected cost of typical process plants
Price Index Numbers for Current Cost Accounting	Indices for nonresidential construction, machinery, equipment
France	
Index Coefficients	General index for building and machinery replacement costs; French

Germany
Fachserie 17: Preise — Indices for general machinery and building construction; German

Italy
Indicatori Mensili — Government statistics and indices; Italian
Index — Costs and indices for various industries; Italian
Prezzi Informativi Delle Opere Edili in Milano — Indices for residential and non-residential construction, consumer prices, raw materials and wage rates; Italian

Japan
Construction Price Indices by Year, Price Indices of General Machinery and Equipment — Non-Life Insurance Institute publication for use by industrial insurers in Japan; English
MRC Monthly Standard Building Cost Indexes and Unit Price Data Bulletin — Cost indices of many components of construction and certain building types; Japanese

Mexico
Cifras de la Construccion — General construction index; Spanish
Indice Nacional de Precios al Consumidor — Consumer price index; Spanish

The Netherlands
Die Werkgroep Begrotings Problemen in de Chemische Industrie (WEBCI) — Unit prices for chemical plant construction in the Netherlands; Dutch

New Zealand
Ministry of Works and Development Construction Cost Index — Weighted construction cost index
Monthly Abstract of Statistics — Variety of labor force indices by industry; residential construction, wage rates, other

United States
General Purpose Indexes — *ENR* 20-city Construction Cost
ENR 20-city Building Cost
U.S. Commerce Dept. composite

(continued)

Table 7.8 *Continued*

Country/Index	Comments
	BuRec, general building
	Construction Industry Institute, const. price
	Factory Mutual, industrial building
	Handy Whitman, building construction
	Lee Saylor, Inc., material/labor
	R. S. Means construction cost
Selling Price Indexes, Building	
	Fru-Con Corp, industrial
	Lee Saylor, Inc. subcontractor
	Turner, general building
	Smith, Hinchman & Grylls, general
Valuation Indexes	
	Boeckh, 20-city commercial/manufacturing
	Marshall & Swift, industrial equipment
Special Purpose	
	Nelson-Farrar refinery cost inflation index
	Chemical Engineering plant cost
	Federal Highway Construction Bid Price
	Handy Whitman Public Utility Construction
	BuLabor Statistics Consumers Price Index
	BuLabor Statistics Producers Price Index
	AED Average Rental Rates for Const. Equipment

[a]For source information and a more detailed review of each of these indexes, see AACE Cost Index Committee (1989).
Source: B. G. McMillan and K. K. Humphreys, "Sources of International Cost Data," *AACE Transactions,* AACE International, Morgantown, WV, 1990, Paper K.6.

Table 7.9 2003 Member Societies of the International Cost Engineering Council (ICEC) and the International Project Management Association (IPMA)

Society	Country	ICEC Member	IPMA Member
AACE-CANADA, The Canadian Association of Cost Engineers	Canada	Yes	No
AACE International, the Association for the Advancement of Cost Engineering (AACE-I) (chapters in Canada, Saudi Arabia, Japan, Australia, South Africa, Trinidad and Tobago, Puerto Rico, Egypt, Malaysia and Norway)	United States	Yes	No
Asociación Espanola de Ingenieria de Proyectos (AEIPRO)	Spain	Yes	Yes
Associacao Portuguesa de Gestao de Projectos (APOGEP)	Portugal	No	Yes
Association of Cost Engineers (ACostE) (chapters in Hong Kong and Siberia)	United Kingdom	Yes	No
Association of Cost Engineers, Hong Kong Branch (ACostE HK)	Hong Kong	Pending	No
Association Françophone de Management de Projet (AFITEP) (chapters in Belgium and Switzerland)	France	Yes	Yes
Association of Management, Consulting and Technology in Construction (CMCTC)	Romania	Yes	No
Association of Project Managers (APM)	United Kingdom	No	Yes
Association of Project Managers of South Africa (APMSA)	South Africa	No	Yes
Association of South African Quantity Surveyors (ASAQS)	South Africa	Yes	No
Associazione Nazionale di Implantistica Industriale (ANIMP)	Italy	No	Yes

(continued)

Table 7.9   *Continued*

Society	Country	ICEC Member	IPMA Member
Associazione Italiana d'Ingegneria Economica (AICE)	Italy	Yes	No
Australian Institute of Quantity Surveyors (AIQS)	Australia	Yes	No
Azerbaijan Project Management Association (AzPMA)	Azerbaijan	No	Yes
Building Suryeyor's Institute of Japan (BSIJ)	Japan	Yes	No
Canadian Institute of Quantity Surveyors (CIQS)	Canada	Yes	No
China Engineering Cost Association (CECA)	China	Pending	Yes
Cost Engineering Association of Southern Africa (CEASA)	11 countries in Southern Africa	Yes	No
Croatian Association for Project Management	Croatia	No	Yes
Cyprus Association of Professional Quantity Surveyors (CAPQS)	Cyprus	Yes	No
Czech Association of Project Management (SPR)	Czech Republic	Yes	Yes
Deutsche Gesellschaft für Projektmanagement eV (GPM)	Germany	No	Yes
Fiji Institute of Quantity Surveyors (FIQS)	Fiji	Yes	No
Foreningen for Dansk Projektledelse (FDP)	Denmark	Yes	Yes
Gépipari Tudományos Egyesület/ Mûszaki Költségtervez Klub (GTE/MKK) (Hungarian Cost Engineering Club)	Hungary	Yes	No
Ghana Institution of Surveyors (GIS)	Ghana	Yes	No
Grupo OFC Guia Referencial de Costos (GOGRC)	Venezuela	Yes	No
Hellenic Project Management Association (HPMI)	Greece	Yes	Yes
Hong Kong Institute of Surveyors (HKIS)	Hong Kong	Yes	No

(*continued*)

Table 7.9   *Continued*

Society	Country	ICEC Member	IPMA Member
Institute of Project Management of Ireland	Ireland	No	Yes
Institute of Project Managers of Sri Lanka (IPMSL)	Sri Lanka	Yes	No
Institute of Quantity Surveyors of Kenya (IQSK)	Kenya	Yes	No
Institute of Namibian Quantity Surveyors (INQS)	Namibia	Yes	No
Institute of Quantity Surveyors Sri Lanka (IQSSL)	Sri Lanka	Yes	No
Instituto Brasileiro de Engenharia de Custos (IBEC)	Brazil	Yes	No
Institution of Surveyors, Malaysia (ISM)	Malaysia	Yes	No
Institution of Surveyors of Uganda (ISU)	Uganda	Pending	No
Japan Society of Cost and Project Engineers (JSCPE)	Japan	Yes	No
Korean Institute of Project Management and Technology (PROMAT)	South Korea	Former Member	No
Latvian National Association of Project Management	Latvia	No	Yes
Management Engineering Society (MES)	Egypt	No	Yes
Mauritius Association of Quantity Surveyors (MAQS)	Mauritius	Yes	No
Nederlandse Stichting Voor Kostentechniek (DACE)	Netherlands	Yes	No
New Zealand Institute of Quantity Surveyors (NZIQS)	New Zealand	Yes	No
Nigerian Institute of Quantity Surveyors (NIQS)	Nigeria	Yes	No
Norsk Forening for Prosjektledelse (NFP)	Norway	Yes	Yes
Project Management Associates	India	Yes	Yes
Project Management Association of Finland (PMAF)	Finland	Yes	Yes
Project Management Association of Hungary (FOVOSZ)	Hungary	No	Yes

(continued)

Table 7.9 *Continued*

Society	Country	ICEC Member	IPMA Member
Project Management Association of Slovakia (SPPR)	Slovakia	Yes	Yes
Project Management Institut Nederland	Netherlands	No	Yes
Project Management - South Africa (PMSA)	South Africa	Yes	No
Project Management Research Committee, China	China	No	Yes
Project Management Romania	Romania	No	Yes
Projekt Management Austria-Institut	Austria	Yes	Yes
Russian Association of Bidding and Cost Engineering	Russia	Yes	No
Russian Project Management Association (SOVNET)	Russia	No	Yes
Schweizerische Gesellschaft für Projektmanagement (SPMS)	Switzerland	No	Yes
Singapore Institute of Surveyors and Valuers	Singapore	Yes	No
Slovenian Project Management Association (ZPM)	Slovenia	Yes	Yes
Sociedad Mexicana de Ingeniería Económica, Financiera y de Costos (SMIEFC)	Mexico	Yes	No
Stowarzyszenie Project Management Polska (SPMP)	Poland	No	Yes
Svenskt ProjektForum (SPMS)	Sweden	Yes	Yes
Tanzania Institute of Quantity Surveyors(TIQS)	Tanzania	Pending	No
Ukranian Project Management Association (UKRNET)	Ukraine	No	Yes
Verkefnastjórnunarfélag Íslands (VSF)	Iceland	Yes	Yes
Yugoslav Project Management Association (YUPMA)	Yugoslavia	No	Yes

ICEC Secretariat address: PO Box 301, Deakin West, ACT 2600, Australia, Phone: 61-2-6282-2222, Fax: 61-2-6285-2427, E-mail: icec@icoste.org, Internet: http://www.icoste.org.
IPMA Secretariat address: PO Box 1167, 3860 BD Nijkerk, The Netherlands, Phone: 31-33-247-34-30, Fax: 31-33-246-04-70, E-mail: info@ipma.ch, Internet: http://www.ipma.ch.

7.2.3 Country Cost Indexes

Cost indexes are valuable proven tools for adjusting costs for changes over time and, in combination with appropriate location factors, can facilitate development of conceptual estimates. The AACE International *Cost Engineers' Notebook* (AACE Cost Index Committee, 1989). describes 28 indexes and index sources for US and Canadian costs and 27 indexes and index sources for many other countries, including Australia, Brazil, England, France, Italy, Japan, Mexico, the Netherlands, New Zealand, and West Germany. A description of each index or index source is given in Table 7.8.

7.2.4 Cost Data Sources

Various compilers and publishers maintain databases of costs which form the basis for their various cost publications. A few of these are listed in Table 7.6.

Beyond these, invaluable sources of information and costs are the various cost engineering and project management societies throughout the world. Their members and employing firms are the most valuable network of international cost information.

At the time of this writing, the 44 member societies of the International Cost Engineering Council (ICEC) represented, either directly or through their branches and chapters, over 40 countries.

The International Project Management Association (IPMA, formerly INTERNET) similarly had about 33 member societies representing many nations. A list of the members of ICEC and IPMA is given in Table 7.9. Internet web page addresses for ICEC and IPMA and for their member societies are provided in Appendix C. However, because some of these societies do not have a web site or a permanent headquarters offices, it is suggested that, in those cases, ICEC or IPMA be contacted for the current addresses of each society.

REFERENCES

AACE Cost Index Committee (1989). Cost/price indexes. *Cost Engineers' Notebook.* Morgantown, WV: AACE International.

ACostE (1982). *Estimating Check List for Offshore Projects.* Sandbach, Cheshire, UK: Association of Cost Engineers.

Bent, J. A. (1996). Humphreys, K. K., ed. *Effective Project Management Through Applied Cost and Schedule Control.* New York: Marcel Dekker.

Bridgewater, A. V. (1979). International construction cost location factors. *Chem. Eng.* 86:5.

Fong, C. K., ed. (Undated). *Report on the Cost Competitiveness of the Construction Industry in Singapore.* Construction Industry Development Board, 133 Cecil Street #09-01/02, Keck Seng Tower, Singapore 0106.

Massa, R. V. (1984). International composite cost location factors. *AACE Transactions.* Morgantown, WV: AACE International.

Massa, R. V. (1985). A survey of international cost indexes, a north American perspective. 1st European Cost Engineering Forum, International Cost Engineering Symposium, Oslo, Norway.

Patrascu, A. (1979). Pre-estimating survey. The international aspect. *AACE Transactions.* Morgantown, WV: AACE International.

Patrascu, A. (1988). *Construction Cost Engineering Handbook.* New York: Marcel Dekker, Inc.

Shellenberger, D. J. (1967). A comparison of building cost indexes. *Cost Engineers' Notebook.* Morgantown, WV: AACE International.

Walker, C. G. (1978). Estimating construction costs abroad. *Transactions Fifth International Cost Engineering Congress.* Dutch Association of Cost Engineers, The Hague, pp. 40–46.

RECOMMENDED READING AND OTHER INFORMATION SOURCES

ACostE Indexes of Erected Plant Costs, Association of Cost Engineers, Ltd., Lea House, 5 Middlewich Road, Sandbach, Cheshire, CW11 9XL United Kingdom.

Associated Equipment Distributors (AED), 615 West 22nd Street, Oak Brook, IL 60523.

Building Economist, The Australian Institute of Quantity Surveyors, PO Box 301, Deakin West, ACT 2600, Australia.

Chemical Engineering, Chemical Week Publishing, L.L.C., 110 William St., 11th Floor, New York, NY 10038.

Construction Industry Institute, 3925 West Braker Lane, Austin, TX 78759-5316.

Fachserie 17: Preise, Verlag W. Kohihammer GmbH, Abt. Veroffentlichungen des Statistischen Bunde Samtes, Philipp-Reis-Strasse 3, Postfach 42 11 20, D-6500 Mainz 42, Hechtsheim, Germany.

Hackney, J. W. (1997). Humphreys, K. K., ed. *Control and Management of Capital Projects.* 2nd ed. Morgantown, WV: AACE International.

Humphreys, K. K., ed. (1991). *Jelen's Cost and Optimization Engineering.* 3rd ed. New York: McGraw Hill Book Company.

Index Coefficients, Roux S.A., 51 Ampere, 75017 Paris, France.

Indicatori Mensili, Instituto Centrale di Statistica, Via Cesare Balbo 16-00100, Roma, Italy.

Japan Construction Price Indexes by Year, Price Indexes of General Machinery and Equipment, The Non-Life Insurance Institute of Japan, 6–5, 3-Chome, Kanda, Surugadai, Chiyoda-ku, Tokyo, Japan.

Marshall & Swift/Boeckh, 2885 S. Calhoun Road, New Berlin, WI 53151.

Miller, C. A. (1978). Selection of a cost index. *Cost Engineers' Notebook.* Morgantown, WV: AACE International.

Ministry of Works and Development Construction Cost Index, Ministry of Works & Development, PO Box 12 041, Wellington North, New Zealand.

Monthly Abstract of Statistics. Department of Statistics, Aorangi House, 85 Molesworth Street, Wellington, New Zealand.

MRC Monthly Standard Building Cost Indexes and Unit Price Data Bulletin, Management Research Society for Construction Industry, Japan.

Price Index Numbers for Current Cost Accounting, Department of Industry, Economics and Statistics, Div. 4A, Sanctuary Buildings, 16–20 Great Smith Street, London SW1P 3D8, England.

R. S. Means Co., 100 Construction Plaza, Kingston, MA 02364.

Statistics Canada, Prices Division, Ottawa, Ontario KlA 0T6 Canada.

Turner Construction Co. 375 Hudson Street, New York, NY 10014.

U.S. Bureau of the Census, Construction Statistics Division, U.S. Dept. of Commerce, Washington, DC 20233.

U.S. Dept. of Labor, Bureau of Labor Statistics (BLS), Washington, DC 20212.

U.S. Dept. of Transportation, Federal Highway Administration, Interstate Management Branch, Washington, DC 20590.

8

Operations Research Techniques

Operations research may be generally described as a scientific approach to decision-making that involves the operations of organizational systems. It is widely used in industry. The most common and useful techniques are linear programming, dynamic programming, game theory, network analysis including PERT (Program Evaluation and Review Technique) and CPM (Critical Path Method), queuing theory, and simulation techniques.

Operations research involves many problems in optimization and thus has wide application in cost engineering, where minimum cost or maximum profit are prime considerations. This chapter will aid the reader in understanding the techniques used in linear programming, dynamic programming, and simulation. All are very useful in the management science aspects of cost engineering.

8.1 LINEAR PROGRAMMING

Linear programming (LP) deals with problems of allocating limited resources among competing activities in some optimal manner. All the mathematical functions in the LP model must be linear.

Linear programming is concerned with a solution (where a solution exists) to problems of the form:

Optimize (i.e., maximize or minimize)

$$z = k_1 x_1 + k_2 x_2 + \cdots + k_j x_j \tag{8.1}$$

subject to the constraints

1.
$$a_{11}x_1 + a_{12}x_2 + a_{13}x_3 + \cdots + a_{1j}x_j \geq a_{10}$$
$$a_{21}x_1 + a_{22}x_2 + a_{23}x_3 + \cdots + a_{2j}x_j \geq a_{20}$$

$$\vdots$$

$$a_{A1}x_1 + a_{A2}x_2 + a_{A3}x_3 + \cdots + a_{Aj}x_j \geq a_{A0}$$

2.

$$b_{11}x_1 + b_{12}x_2 + b_{13}x_3 + \cdots + b_{1j}x_j \leq b_{10}$$
$$b_{21}x_1 + b_{22}x_2 + b_{23}x_3 + \cdots + b_{2j}x_j \leq b_{20}$$
$$\vdots$$
$$b_{B1}x_1 + b_{B2}x_2 + b_{B3}x_3 + \cdots + b_{Bj}x_j \leq b_{B0}$$

3.

$$c_{11}x_1 + c_{12}x_2 + c_{13}x_3 + \cdots + c_{1j}x_j = c_{10}$$
$$c_{21}x_1 + c_{22}x_2 + c_{23}x_3 + \cdots + c_{2j}x_j = c_{20}$$
$$\vdots$$
$$c_{C1}x_1 + c_{C2}x_2 + c_{C3}x_3 + \cdots + c_{Cj}x_j = c_{C0}$$

Thus we have a *linear* objective function which is to be optimized subject to a total of $A + B + C$ linear constraints involving inequalities ($\geq$ and $\leq$) and equalities.

One key to the successful utilization of linear programming technology is the ability to recognize when a problem can be solved by linear techniques and to properly formulate the mathematical model. Let us examine three linear programming applications.

Manufacturing Problem

A manufacturing firm has excess capacity to devote to one or more of three products called 1, 2, and 3. The excess capacity is

Milling machine	200 hr/wk
Lathe	100 hr/wk
Grinder	50 hr/wk

The manufacture of the three products requires the following machine time (in hours per unit product):

	Milling machine	Lathe	Grinder
Product 1	8	4	2
Product 2	2	3	0
Product 3	3	0	1

The sales potential for products 1 and 2 is estimated to exceed the maximum production rate, and for product 3 it is estimated to be 20 units per week. The unit profit would be $20, $6, and $8, respectively, on products 1, 2, and 3. What production schedule will maximize the profit?

Let $x_1(i = 1, 2, 3)$ be the number of units of product i produced per week. The measure of effectiveness is the profit, denoted by z; we wish to maximize it. The mathematical formulation is:

Maximize

$$z = 20x_1 + 6x_2 + 8x_3$$

subject to the restrictions

Machine capacity

$$8x_1 + 2x_2 + 3x_3 \leq 200$$
$$4x_1 + 3x_2 \qquad \leq 100$$
$$2x_1 + \qquad x_3 \leq 50$$
$$x_3 \leq 20$$

Sales potential nonnegativity:

$$x_1, x_2, x_3 \leq 0$$

The nonnegativity requirement prevents the mathematically possible, but realistically impossible, solution of negative production.

Product Mix Problem

A refinery wishes to blend four petroleum constituents, 1, 2, 3, and 4, into three grades of gasoline, called products A, B, and C. The constituent availability (bbl/day) and cost per barrel are:

Constituent 1: 3000 at $3
Constituent 2: 2000 at $6
Constituent 3: 4000 at $4
Constituent 4: 1000 at $5

The mixtures of the gasoline products are:

Product A: $\leq$30% of 1, $\geq$40% of 2, and $\leq$50% of 3
Product B: $\leq$50% of 1 and $\geq$10% of 2
Product C: $\leq$70% of 3

The selling price per barrel is $5.50, $4.50, and $3.50, respectively, for products A, B, and C. If there are no sales potential or equipment availability considerations, what product mix will maximize the profit?

Let x_{A1} be the amount of Constituent 1 allocated to Product A, x_{B2} be the amount of Constituent 2 allocated to Product B, and so on. Denoting the profit by z, we wish to maximize

$$z = (\text{total sales}) - (\text{total costs})$$
$$= 5.5(x_{A1} + x_{A2} + x_{A3} + x_{A4})$$
$$+ 4.5(x_{B1} + x_{B2} + x_{B3} + x_{B4})$$
$$+ 3.5(x_{C1} + x_{C2} + x_{C3} + x_{C4}) - 3(x_{A1} + x_{B1} + x_{C1})$$
$$- 6(x_{A2} + x_{B2} + x_{C2}) - 4(x_{A3} + x_{B3} + x_{C3})$$
$$- 5(x_{A4} + x_{B4} + x_{C4})$$

subject to the following restrictions
    Availability:

$$x_{A1} + x_{B1} + x_{C1} \leq 3000$$
$$x_{A2} + x_{B2} + x_{C2} \leq 2000$$
$$x_{A3} + x_{B3} + x_{C3} \leq 4000$$
$$x_{A4} + x_{B4} + x_{C4} \leq 1000$$

Blend Restrictions:
    For Product A:

$$x_{A1} \leq 0.3(x_{A1} + x_{A2} + x_{A3} + x_{A4})$$
$$x_{A2} \geq 0.4(x_{A1} + x_{A2} + x_{A3} + x_{A4})$$
$$x_{A3} \leq 0.5(x_{A1} + x_{A2} + x_{A3} + x_{A4})$$

For Product B:

$$x_{B1} \leq 0.5(x_{B1} + x_{B2} + x_{B3} + x_{B4})$$
$$x_{B2} \geq 0.1(x_{B1} + x_{B2} + x_{B3} + x_{B4})$$

For Product C:

$$x_{C1} \leq 0.7(x_{C1} + x_{C2} + x_{C3} + x_{C4})$$

Nonnegativity:

$$x_{A1}, \ldots, x_{A4}, x_{B1}, \ldots, x_{B4}, x_{C1}, \ldots, x_{C4} \geq 0$$

Investment Problem

An investor has $10,000 which can be invested in money-making activities A, B, C, and D. The money-making activities become available as follows:

A and B:	immediately
C:	beginning of second year
D:	beginning of fifth year

The profit payouts for each dollar invested in a money-making activity are as follows:

A:	$0.40 after 2 years
B	$0.70 after 3 years
C:	$1.00 after 4 years
D:	$0.30 after 1 year

Assuming that each investment can be immediately renewed after a payout period, how should the $10,000 be invested so as to maximize the total accumulated payout at the end of 5 years?

Let z represent the total payout after 5 years; and let x_{A1} represent the amount in dollars invested in Activity A the first year, and so on. The problem is then to maximize

$$z = 0.4(x_{A1} + x_{A2} + x_{A3} + x_{A4}) + 0.7(x_{B1} + x_{B2} + x_{B3}) + 1.0x_{C2} + 0.3x_{D5}$$

subject to the restrictions

Year 1: $x_{A1} + x_{B1} \leq 10,000$

Year 2: $x_{A1} + x_{B1} + x_{A2} + x_{B2} + x_{C2} \leq 10,000$

Year 3: $x_{B1} + x_{A2} + x_{B2} + x_{C2} + x_{A3} + x_{B3} \leq 10,000 + 0.4x_{A1}$

Year 4: $x_{B2} + x_{C2} + x_{A3} + x_{B3} + x_{A4} \leq 10,000 + 0.4(x_{A1} + x_{A2}) + 0.7x_{B1}$

Year 5: $x_{C2} + x_{B3} + x_{A4} + x_{D5} \leq 10,000 + 0.4(x_{A1} + x_{A2} + x_{A3})$
$$+ 0.7(x_{B1} + x_{B2})$$

Nonnegativity: $_{A1}, \ldots, x_{A4}, x_{B1}, \ldots, x_{B3}, x_{C2}, x_{D5} \leq 0$

Each of the preceding sample problems contained three or more variables. Where no more than three variables are involved, a graphical solution can be obtained. A two-variable problem involves a two-dimensional display and can best be utilized to demonstrate the technique.

8.1.1 Two-Dimensional Graphical Solution

The profits from Products 1 and 2 are $1 and $2, respectively, per unit. Assuming that x_1 and x_2 units of the products are to be produced, we wish to maximize the profit

$$z = 1(x_1) + 2(x_2)$$

subject to the restrictions

1. $-x_1 + 3x_2 \leq 10,000$
2. $x_1 + x_2 \leq 6,000$
3. $x_1 - x_2 \leq 2,000$
4. $x_1 + 3x_2 \geq 6,000$

$$x_1, x_2 \geq 0$$

Figure 8.1 shows pictorially the restricted solution area and the straight-line loci for the two values of the profit within the area. We note first that the

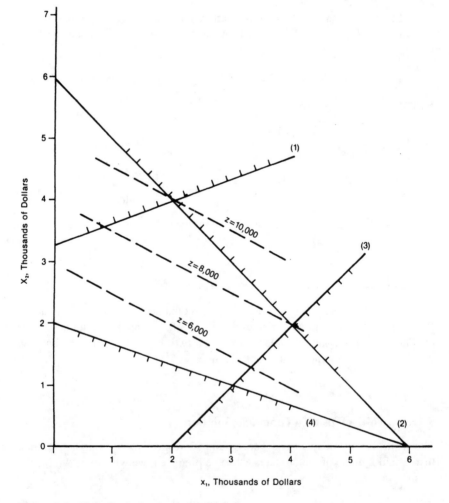

Figure 8.1 Two-dimensional graphical solutions.

nonnegativity constraint eliminates all but the first quadrant from consideration. Next we note that as the profit z increases, the loci of constant z values are parallel, with a slope -0.5. Clearly the maximum value of z that satisfies the restrictions occurs at the intersection of restriction boundaries (1) and (2); it is $z = 10,000$ for which $x_1 = 2000$ and $x_2 = 4000$.

The solution is unique, and it is finite. Had restriction (2) not been imposed, the solution would have occurred at the intersection of restriction boundaries (1) and (3). Had both restrictions (1) and (2) not been imposed, there would have been no finite solution for z. Had the profit been given by $z = x_1 + x_2$, we see that there would have been a finite line of solutions along restriction boundary (2) between its intersections with restriction boundaries (1) and (3).

8.1.2   Simplex Method

Any value $(x_1, x_2, \ldots, x_N)$ of the linear programming variables for which all the restrictions are satisfied is called a *feasible point* or *solution*. The function z to be optimized is called the *objective function*, and a feasible point that maximizes it is called an *optimal solution*. The intersections of the restriction boundaries on the perimeter of the feasible solution region are called *extreme feasible points* or *solutions*. The optimal solution can be shown to always be an extreme feasible point because of the linearity of the objective function and the restrictions; in the case of two extreme points being solutions, the optimal solution is "nonunique," and it consists of the entire edge joining the points.

Since the solution occurs at an extreme feasible point, we may begin at some such point and compute the value of the objective function z. We then exchange this extreme point for its counterpoint at the other end of an edge in such a manner that a larger value of z is obtained; keep in mind that an increasing number of edges will intersect at any extreme feasible point as the dimension of the linear programming problem increases. The process of exchange and edge following continues until z can no longer be increased, at which point (or points) we have one or more optimal solutions. In this manner all of the feasible points are not examined; their number can be prohibitive for large problems. Furthermore, since each successive extreme point must increase the value of z, it is impossible to return to an extreme point previously reached. A well-known algorithm for effecting this procedure, the *simplex* method, was illustrated in Chapter 1.

It is beyond the scope of this chapter to investigate the details of the simplex algorithm. Rather, we will state precisely the form of the problem that the method will solve, and concentrate on converting linear programming problems into that form. Problems can then easily be formulated for, and solved by, computer programs.

The simplex algorithm solves problems of the form:
Maximize

$$z = c_1 x_1 + c_2 x_2 + \cdots + c_N x_N \tag{8.2}$$

subject to the constraints

$$a_{11}x_1 + a_{12}x_2 + \cdots + a_{1N}x_N + x_{N+1} = b_1$$
$$a_{21}x_1 + a_{22}x_2 + \cdots + a_{2N}x_N + 0 + x_{N+2} = b_2$$

$$\vdots$$

$$a_{M1}x_1 + a_{M2}x_2 + \cdots + a_{MN}x_N + 0 + 0 + \cdots + x_{N+M} = b_M$$
$$x_1, x_2, \ldots, x_N, b_1, b_2, \ldots, b_M \geq 0$$

where n = number of real variables and m = number of constraints

Note that each of the variables x_{N+1} through x_{N+M} occurs uniquely in a single constraint, each having a coefficient value of 1. These are termed the *initial basis variables*.

This is a significant difference from the previous problems, notably expression (8.1), with which strictly linear programming is concerned. We need to examine how to determine if a given problem is indeed convertible to this simplex form.

8.1.3 Conversion to Simplex Form

The conversion to the simplex form is accomplished by the introduction of "slack," "surplus," and "artificial" variables, in addition to some mathematical finagling. We have the following definitions:

> *Slack variable*: the positive amount added to the left-hand side of a " $\leq$ " constraint to give an equality " $=$ "
>
> *Surplus variable*: the positive amount subtracted from the left-hand side of a " $\geq$ " constraint to give an equality " $=$ "
>
> *Artificial variable*: a variable that is introduced into the linear programming problem to aid in the conversion to the simplex form, but which will be forced to be zero in the optimal solution

Let us consider the conversion of a simple linear programming problem to the simplex form through the utilization of only slack and surplus variables.

Original form

Maximize: $z = 2x_1 + x_2$

subject to the restrictions

1. $x_1 + x_2 \geq 6$
2. $x_1 + 3x_2 \geq -5$
   $x_1, x_2 \geq 0$

Simplex form

Maximize: $z = 2x_1 + x_2 + 0(x_3 + x_4)$

1. $x_1 + x_2 + x_3 \qquad = 6$
2. $-x_1 - 3x_2 \qquad + x_4 = 5$
$$x_1, \ldots, x_4 \geq 0$$

Here we have introduced an as-yet-unknown slack variable x_3, which simply takes up the slack between the two sides of the "$\leq$" sign, and an as-yet-unknown surplus variable x_4. It is clear that as long as the slack variable satisfies the single restriction $x_3 \geq 0$, then $(x_1 + x_2) \leq 6$ in the equation $(x_1 + x_2) + x_3 = 6$. A similar argument applies to restriction 2. Upon examining the basic simplex form of the linear programming problem, we see that x_3 and x_4 have become the initial basis variables and that the right-hand constants of the simplex form of the restrictions are positive, as they must be.

The example above was quite simple because there were no complicating factors in its conversion. Let us now consider four common complications and their solutions.

Nonpositive b_i: Had constraint 2 of the preceding problem been $x_1 + 3x_2 \geq 5$, the simple introduction of the surplus variable x_4 leads to the equivalent simplex expression

$$-x_1 - 3x_2 + x_4 = -5.$$

To make the right-hand constant positive, we can rewrite the expression as

$$x_1 + 3x_2 - x_4 = 5$$

but now x_4 has a -1.0 coefficient and does not meet the criteria for a basis variable. We can introduce a fifth positive variable x_5 to be added to the left-hand side of this expression which will serve as an initial basis variable for the expression, but the original inequality (constraint 2) will no longer be represented by the expression unless x_5 is forced to zero. We can drive this "artificial" variable to zero by including it in the original objective function with a large negative coefficient, say $(-M)$. The simplex formulation of the entire sample problem then becomes:

Maximize

$$z = 2x_1 + x_2 - Mx_5 \quad (M \gg 0)$$

subject to the restrictions

1. $x_1 + x_2 + x_3 \qquad = 6$
2. $x_1 + 3x_2 \qquad - x_4 + x_5 = 5$
$$x_1, \ldots, x_5 \geq 0$$

We see that since $x_5 \geq 0$, the objective function can only be maximized when $x_5 = 0$. This is commonly called the "big M" method.

Equality in the restrictions: The problem here is that there is no evident means to introduce a slack or surplus variable in order that an equality contain an initial basis variable. One method for alleviating this problem is to replace an equality by a pair of inequalities; for example,

$$3x_1 + 2x_2 = 18$$

is equivalent to the pair of inequalities

$$3x_1 + 2x_2 \leq 18$$
$$3x_1 + 2x_2 \geq 18$$

Where there is more than one equality, we can add all the expressions together and then convert the composite expression into a pair of inequalities. Perhaps the simplest and most desirable technique for alleviating the equality restriction problem is to introduce an artificial variable in the left-hand side of the equality, give it a large negative coefficient, and include it in the objective function (i.e., we apply the big M method).

Variables unconstrained in sign: We note that all variables in the simplex form of the linear programming problem are constrained to be ≥ 0. Certainly many realistic linear programming problems can be developed in which the variables must be negative in the optimum solution. We solve the problem in such cases by merely replacing each unconstrained variable, say x_i, by the simple difference in two positive variables. Thus we make the substitution

$$x_i = x_i' - x_j'$$

$$x_i, x_j \geq 0$$

where x_j' is a new variable (i.e., the subscript j has not been previously assigned in the problem). We must then remember to convert back to the original variable x_i by subtracting x_j from it x_i' in the optimum solution. This procedure is implemented independently for each such unconstrained variable.

Objective function to be minimized: Here we simply utilize the principle that minimizing any function $f(x_1, x_2, \ldots, x_n)$ subject to a set of restrictions is completely equivalent to maximizing $-f(x_1, x_2, \ldots, x_n)$ subject to the same restrictions. We then take the negative of the optimum solution for the latter objective function to obtain the minimized result.

Let us consider the conversion of a second linear programming problem to the simplex form, where all four complications exist. The original form of the problem is:

Minimize

$$z = 2x_1 + x_2 - 6x_3 - x_4$$

subject to the restrictions

1. $3x_1 + \qquad\qquad x_4 \le 25$
2. $x_1 + x_2 + x_3 + x_4 = 20$
3. $4x_1 + \qquad 6x_3 \qquad \ge 5$
4. $2 \le 2x_2 + 3x_3 + 2x_4 \le 30$
5. $2x_1 + x_2 - 6x_3 - x_4 \le 100$
$$x_1, \ldots, x_3 \ge 0$$

The simplex form of the problem is:
Maximize

$$(-z) = -[2x_1 + x_2 - 6x_3 - (x_4' - x_5')] - M(x_7 + x_9 + x_{11})$$

subject to the restrictions

1. $3x_1 \qquad\qquad + (x_4' - x_5') + x_6 \qquad\qquad\qquad\qquad = 25$
2. $x_1 + x_2 + x_3 + (x_4' - x_5') \qquad + x_7 \qquad\qquad\qquad = 20$
3. $4x_1 \qquad + 6x_3 \qquad\qquad\qquad - x_8 + x_9 \qquad\qquad = 5$
4. $\qquad 2x_2 + 3x_3 + 2(x_4' - x_5') \qquad\qquad - x_{10} + x_{11} = 2$
4'. $\qquad 2x_2 + 3x_3 + 2(x_4' - x_5')$
$\qquad\qquad\qquad + x_{12} \qquad\qquad\qquad\qquad\qquad\qquad = 30$
5. $2x_1 + x_2 - 6x_3 - (x_4' - x_5')$
$\qquad\qquad\qquad + x_{13} \qquad\qquad\qquad\qquad\qquad\qquad = 100$

$$x_1, \ldots, x_3, x_4', x_5', x_6, \ldots, x_{13} \ge 0$$

The solution to the simplex formulation of the problem is

$$(x_1, \ldots, x_3, x_4', x_5', x_6, \ldots, x_{13}; -z)$$
$$= (20, 5, 30, 0, 35, 0, 0, 255, 0, 28, 0, 0, 0, 0;\ 100)$$

We now must make the conversion $x_4 = x_4' - x_5'$ and take the negative of the objective function to obtain the solution to the original problem; we obtain

$$(x_1, \ldots, x_4;\ z) = (20, 5, 30, -35;\ -100)$$

where $(z = 2x_1 + x_2 - 6x_3 - x_4)$ has been minimized, subject to the specified restrictions.

8.1.4 Interpretation of Simplex Solutions

There are several outcomes to a simplex solution of a linear programming problem:

1. A unique solution is obtained.
2. Multiple solutions are obtained (nonunique solution).
3. No feasible solution exists.
4. No bounded solution exists.

A discussion of the last three outcomes follows.

Multiple solutions: There are no means for analyzing the uniqueness of a single simplex solution. If there are multiple solutions, the determination of all solutions requires a special algorithm that is extremely sensitive to computer roundoff error. However, the simplex problem can be reformulated by reversing the order to the subscripts of the variables in the original objective function. A different solution should then be obtained if the original solution was not unique. Where there are multiple solutions, an infinite number of optimal solutions can be constructed by taking weighted averages of the simplex solutions; thus, if the two solutions $(x_1, \ x_2) = (2, 3)$ and (3, 4) are obtained, other optimal solutions are $(x_1, x_2) = (2/2 + 3/2, 3/2 + 4/2), (2/3 + 6/3, 3/3 + 8/3)$, and so on.

No feasible solution: If one of the artificial variables has a value other than zero in the final solution, no feasible solution exists for that particular simplex formulation. The coefficient M of the artificial variables in the objective function can be increased (provided that roundoff error is not prohibitive) and another solution attempted, as a further check on the existence of a feasible solution.

Unbounded solution: Clearly not all linear programming problems will have finite or bounded solutions. There are definite checks for the existence of this situation in the simplex algorithm, and most simplex computer programs will indicate this result when it occurs. An additional feature of the simplex solution is that redundancies in the constraints will result in fewer nonzero variables in the final simplex solution than there are constraints; this occurrence does not, however, ensure that there are redundancies. In such cases there is nothing wrong with the solution; it merely indicates that the original problem was formulated indiscriminately.

8.2   DYNAMIC PROGRAMMING

Dynamic programming is a mathematical procedure often useful for making a sequence of interrelated decisions for a system that is subject to constraints such that some function is optimized by the combination of decisions. In other words, when a task requires multiple, sequential decisions, and each decision impacts the succeeding decision, dynamic programming is a tool to obtain the optimal set of decisions for reaching the desired goal, such as minimized time or expense. (Chapter 1 shows an example of dynamic programming). In contrast to linear programming, a standard mathematical formulation of dynamic programming problems does not exist, although some successful algorithms are used frequently. The best general description of the process is that dynamic programming is a family of algorithms that follows the same general approach to problem solving.

A degree of ingenuity and insight into the general structure of dynamic programming is required to recognize when and how it can be applied, and to which problems. Not all problems are solvable by dynamic programming techniques;

for those that are solvable, not all algorithms apply. It may be tempting to apply a computer program to a dynamic problem, but without understanding the methods the software uses to approach a solution a person may unknowingly generate an inappropriate answer. The only way to learn this technique is to be exposed to a number of problems and their solutions and practice using the technique. One chapter in one book cannot by itself teach the technique. Anyone with a need to learn dynamic programming is well advised to obtain a book on the subject and, best of all, attend a course if one is available.

The expansion of modern techniques has reduced the need of the average professional to learn dynamic programming for several reasons. Specialists handle this task, most corporations have set procedures for various types of analyses, and alternate methods have been developed. It still can be useful, however, to understand the advantages and disadvantages of this technique.

One advantages is that the user has control at any stage of the specific optimization techniques to be used (assuming adequate knowledge of, or flexibility of, the software being applied if computer programs are used). Constraints are an aid rather than the hindrance that they are with other techniques because constraints help reduce the options available and limit the range of the investigation. No analytical function is involved, so discontinuities are easily handled. In general, dynamic programming greatly reduces the number of options to be examined and makes extremely complex problems more approachable.

On the other hand, dynamic programming breaks a problem into stages but does not limit the number of considerations to be taken into account within each stage. This is a problem of dimensionality; the decision at each stage can become exceedingly difficult, and few ways may be available to reduce the options that have to be examined. Also, as in the example given in Chapter 1, each decision taken is independent of all preceding decisions and actions. In reality, this assumption would rarely be valid. In many situations later stages can be affected by the manner in which prior stages occurred; ideal choices may not be selected if these relationships are not considered.

Because of the subjective nature of the technique, even when computerized applications are used, because of dimensionality constraints, and because of the relatively independent treatment of stages, other techniques have become increasingly popular for use in optimization problems. Most notably, stochastic techniques are attractive and useful in applications that were previously impossible to consider. The sheer volume of calculation required to use stochastic methods had made them unattractive for most complex applications. However, the speed and power of modern computers make it possible to perform the millions of calculations necessary to use stochastic analyses effectively.

8.3 SIMULATION

The technique of simulation (often called the "Monte Carlo" technique) has long been an important tool of the designer, whether used to simulate airplane flights in a wind tunnel or lines of communication with an organizational chart. With the advent of the high-speed digital computer with which to conduct simulated experiments, this technique has become increasingly important to operations researchers. Simulation has been said to be the experimental arm of operations research.

The operations research approach to problem solving always involves the formulation and solution of mathematical models that represent real systems. If the model is amenable to solution, an analytical approach usually is superior to simulation. However many problems are so complex that they cannot be solved analytically, and indeed many problems are impossible to describe in mathematical terms. In such cases numerical simulation often affords the only practical approach to a problem. Furthermore, owing to the many developments in digital computer technology, it is often more convenient, inexpensive, and time-saving to solve relatively simple problems by simulation.

Simulation is nothing more or less than the technique of performing sampling experiments on the mathematical model of a system. Thus models "represent" reality and simulation "imitates" it. Simulated experiments should be viewed as virtually indistinguishable from ordinary statistical experiments wherein stochastic processes are an intrinsic reality, and they must be based on sound statistical theory. The heart of the statistical facet of simulation is the "random variate" or "random variable." The illustrative example of a trivial numerical simulation that follows should enhance the understanding of these concepts.

8.3.1   Coin Problem

Suppose that you were offered the chance to play a game whereby you would repeatedly flip an unbiased coin until a sequence of three tails or three heads was tossed. You would receive $8 at the end of each game, and you would be required to pay $1 for each flip of the coin. Assuming that your supply of money is not significantly limited, should you play the game?

The statistical-analytical solution is straightforward but tricky. In any sequence of three throws, the first throw is of no significance, and the probability of repeating the first throw two times in succession, regardless of what the first result was, would be $(1/2)(1/2) = 1/4$. Thus the odds are $4:1$ against you in any sequence of three throws. The odds are then *even* that you will win the game once in a total of two sequences. A total of two sequences costs you a *maximum* of $4, where the sequences are throws (1, 2, 3) and throws (2, 3, 4), an average of $2 per sequence. However additional sequences, when required, cost only $1 each, as each new throw terminates a new sequence of three. Thus the game appears to be a good deal for you, since the return on even odds is even.

A more quantitative analysis of the game can be made by simulation. One technique for simulating the game would be to play the game alone by repeatedly flipping a coin until it becomes clear whether it is worthwhile playing for stakes, a time-consuming task. A better idea would be to simulate the game numerically using either a table of random numbers or a computer-generated sequence of random numbers.

The simulation does not necessarily confirm our suspicions about the game, although it clearly does not contradict them, because there is some finite probability that only 100 simulated flips of the coin will yield erroneous statistics. However, as the number of simulated flips increases, the probability of erroneous results becomes infinitesimal. Indeed, in reality, a computer numerical simulation would be completely automated to the extent that the hand analysis described above would be programmed into the computer, and the only computer output would be the final statistics for literally thousands of simulated coin flips—all simulated in a few microseconds.

For example, assume that the computer can generate a sequence of digits 0, 1, . . . , 9, each having an equal probability of occurrence. Since the probability of a head would be one-half, we assign digits 0, . . . , 4 to the event of a head, and we assign digits 5, . . . , 9 to the event of a tail. A sequence of 100 actual computer-generated digits follows: 8137271655, 7900343568, 5894804865, 3592579729, 3985892576, 9760739827, 1032627137, 0441832139, 5965038789, 5408380131. Converting to heads H and tails T and separating the games by a "/," we obtain the following numerically simulated games: THHTHTHTTT/ TTHHH/HHTTT/TTT/HTHHTTT/HTTHTTT/THTHTTT/TTHTTT/TTT/ HTHTTHTHHH/HTHTHHTHHH/HTHHH/HTTT/HTHHTTT/TTHHTHTH HH/H. Thus we see that 15 games were won, yielding \$120, but the total cost was \$100, yielding 6 : 5 odds of making a profit.

8.3.2 Random Variables

In the coin problem just described we made use of a computer-generated series of random digits with equal probability of occurrence, a seemingly simple task. Tables of random numbers are available. Random numbers can be generated in many pocket programmable calculators. They can, of course, be generated on command in computers. Cost engineers should understand the use of random numbers in simulation, but they need not be experts on the mathematics involved in their generation.

Although random numbers are readily generated with a "flat" distribution, that is, equal probability of occurrence, in many problems it is necessary to generate random variates with numerous probability density functions other than flat. Computers are capable of doing this directly and can generate random variates for the following distributions, among others:

- Exponential distribution
- Poisson distribution
- Chi-square distribution
- Binomial distribution

Chapter 9 discusses the practical use of Monte Carlo simulation in analyzing risk of projects overrunning their estimated cost and in determining the required contingency that should be included in an estimate to reduce the overrun probability to an acceptable level.

8.3.3 Queuing Problem

The multiple facets and applications of numerical simulation make a detailed description of simulation techniques and practices impossible. Perhaps more insight into the workings of numerical simulation can be gained from examination of additional examples. An important field of numerical simulation is queuing theory.

Queuing theory involves the mathematical study of queues or waiting lines. The formation of queues is, of course, a common phenomenon that occurs whenever the current demand for a service exceeds the current capacity to provide that service. The function of queuing theory is to provide realistic mathematical models of queue systems and often to provide analytical solutions to them.

A basic queuing theory model is the "single-server" model with Poisson input and exponential service time. Figure 8.2 depicts the model. The *elapsed time between arrivals* τ has the probability density function

$$f(\tau) = \frac{e^{\left(\frac{\tau}{\tau_m}\right)}}{\tau_m} \tag{8.3}$$

where τ_m is the mean τ; and the service facility requires a time t to perform the service with a probability density function

$$f(t) = \frac{e^{\left(\frac{\tau}{\tau_m}\right)}}{t_m} \tag{8.4}$$

where t_m connotes mean time. The problem is to determine the average waiting time in the queue. The solution of this problem has been determined analytically and is quite well known in queuing theory. The waiting time in the queue W_q is

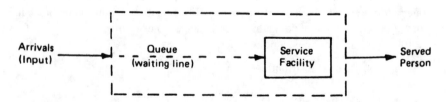

Figure 8.2 Single-server model.

given by

$$W_q = \frac{\left(\dfrac{t_m}{\tau_m}\right)}{\left(\dfrac{1}{t_m} - \dfrac{1}{\tau_m}\right)} \tag{8.5}$$

and the probability density function of waiting time, including service, is known to fit an Erlang distribution curve. For $t_m = 12$ and $\tau_m = 20$ it is found from the above equation that W_q, the expected waiting time in the queue, is 18 time units.

As an example of simulation, the problem has been computer programmed for a solution by asynchronous timing. The procedure is quite simple and is outlined as follows:

1. The simulation is initialized at time zero by entering a person into the queue and then immediately into the service facility, as it is initially empty. At this time the next arrival time and the service time for the initial person are determined.

2. Time is incremented 1 unit, and a test is made to determine (a) if it is time for another person to enter the queue, and/or (b) if it is time for another person to leave the queue and enter the service facility. Then (a) another person is added to the queue if it is time to do so, and (b) if it is time to do so, a person leaves the queue (if any person is waiting there) and enters the service facility. If nobody is available for the service facility, it remains idle for at least 1 unit of time.

3. Time is again incremented 1 unit and step 2 is repeated, and so on, until a prespecified number of persons has entered the queue.

4. In the meantime, at each time increment an account is made of the number of people in the queue, and the total person-time units in the queue is accumulated.

5. The total accumulated person-time units in the queue is then divided by the total number of people who have been serviced to determine the average waiting time in the queue.

   The results from the computer program for $t_m = 12$ and $\tau_m = 20$ are as follows:

Number of arrivals	Mean waiting time
1000	23.13
2000	19.39
3000	19.02
4000	18.04
5000	17.84

The simulation results are seen to range from 23.13 to 17.84 time units, generally increasing in accuracy as the length of the simulation increases.

8.3.4   Discussion

The vast utility of simulation practices can probably be envisaged, even by an amateur, from the few preceding examples. The discussion has been quite general and has not treated any of the specific problems and considerations of simulation. Perhaps the most important of such considerations are the steady-state condition of the simulation, the sample size, the number of samples, and sampling scheme. Too frequently inadequate attention is paid to the sampling scheme, and results are thereby invalidated.

The steady-state condition is a concern where systems being simulated operate continually in a steady-state condition. Unfortunately a simulation model cannot be operated this way; it must be started and stopped. Because of the artificiality introduced by the abrupt beginning of operation, the performance of the simulated system does not become representative of the "real-world" system until it, too, has reached a steady-state condition. Therefore the data obtained during this initial period of operation should be excluded from consideration. Furthermore the simulation should be started in a state that is as representative of steady-state conditions as possible.

The choice of sample interval or size is quite important in planning a simulation study. For example, in the queuing problem, the sample interval was, for all practical purposes, 5% of the average number occurring in the simulation. For the coin problem the sampling interval was essentially infinitesimal and imposed no restrictions on the accuracy of the results. Clearly one's

Table 8.1   Queuing Problem Results

Run	Sample interval (time units)		Run	Sample interval (time units)	
	1	0.1		1	0.1
1	13.18	15.00*	11	16.96*	14.33
2	20.25*	12.90	12	15.38*	15.09
3	15.25*	11.04	13	22.01	14.57*
4	19.36	18.85*	14	11.09	17.22*
5	12.55	21.53*	15	21.61	16.47*
6	12.22	13.75*	16	17.02*	15.65
7	17.25*	20.16	17	17.07*	21.08
8	15.93*	12.17	18	9.92	14.57*
9	10.03	11.17*	19	16.39	18.29*
10	12.30	15.84*	20	15.74*	15.53

Asterisks indicate those values that are closest to the correct value (18.0).

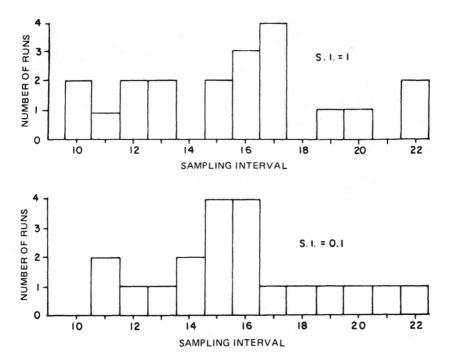

Figure 8.3   Histogram of queuing problem solutions.

confidence in the results of the queuing problem should be enhanced by the use of a smaller sample interval. Table 1 shows the results for the queuing problem where the sample interval was respectively 1 time unit and 0.1 time unit for 20 consecutive independent simulations of 500 samples each. The results for the smaller sample interval are closer overall to being correct; on a run-by-run basis they were closer on 11 of the 20 runs.

The effects of sample interval and sample length on numerical simulation results are essentially indistinguishable, especially when compensation has been made for the initial stabilization period; a decrease in the sample interval increases the simulation calculations by the same amount as a corresponding increase in sample length. A "confidence interval" for some specified sample interval and sample size can be established by the analysis of a sequence of simulations, which provide a sequence of statistically independent observations. As the observations represent averages, they should tend to have an approximately normal distribution as described by the central limit theorem. Furthermore, by making the runs consecutively and by using the ending condition of one run as the steady-state starting condition for the next run, the stabilization period, where it is of concern, can be limited to the single initial run. Approximately 15 to 20 observations should prove to be adequate. The data from Table 8.1 are plotted as histograms in Fig. 8.3. Here we see that the variance of the observations with the smaller sample interval is somewhat smaller; also their mean value is just slightly better (closer to 18.0)—15.8 versus 15.5, although the correct value would not be known for comparison with an actual simulation.

Techniques for sampling so as to decrease the variance of the observations above, resulting in more dependable results, are examples of Monte Carlo techniques (although the term is frequently applied to simulation in general). Since these techniques tend to be rather sophisticated, it is not possible to explore them here. Stratified sampling, importance of sampling, and the control variate method are examples of these techniques.

In conclusion, simulation is indeed a very versatile tool, but by no means is it a panacea. It is invaluable for use on problems where conventional analytical techniques are out of the question, but it does not provide much insight into the cause-and-effect relationships within a system. Simulation requires a large number of arithmetic calculations, but the new generations of computers are so fast and relatively inexpensive that this no longer presents a problem.

RECOMMENDED READING

Humphreys, K. K., ed. (1991). *Jelen's Cost and Optimization Engineering*. 3rd ed. New York: McGraw Hill Book Company.

Rardin, R. L. (1997). *Optimization in Operations Research*. Englewood Cliffs, NJ: Prentice Hall.

Vanderbei, R. J. (2001). *Linear Programming: Foundations and Extensions*. 2nd ed. Boston: Kluwer Academic Publishers.

9

Risk Analysis: How to Do it and How Not to Do it

9.1 INTRODUCTION

Chapter 4 discussed a simplified manual method of assessing project cost risk. While the approach described there has some utility, it is far more practical and in most cases requires less effort to use Monte Carlo software approaches to estimate cost overrun probability and the amount of contingency required to reduce the overrun probability to an acceptable level. The software approaches produce more reliable results than the manual method of Chapter 4. They also enable the cost engineer to explore alternate project scenarios quickly in order to determine the optimum solution for virtually any economic decision, something which is often not practical or feasible with manual methods. The software provides the cost engineer with an estimate of required contingency to avoid overruns at any desired probability level, it identifies the required contingency for each critical project item, and it identifies those items within the estimate that contribute most significantly to project risk, as well as those that afford the greatest areas of opportunity.

Perhaps the most influential pioneer in the field of project risk analysis is Michael Curran, president of Decision Sciences Corporation in St. Louis, Missouri. Curran (1976, 1988, 1989, 1998) has written extensively on project risk analysis and range estimating approaches and developed the first major piece of software for project risk analysis, *Range Estimating Program for Personal Computers* (REP/PC). Curran's pioneering work began in 1964 and was prompted by an article in the *Harvard Business Review*, Jan/Feb 1964 issue entitled "Risk Analysis in Capital Investment" by David B. Hertz.

Curran recognized that, although Hertz never used the term, the article was about Monte Carlo simulation and how it could be used to address key issues in the problem of capital planning. Monte Carlo simulation techniques had come into prominence in the 1940s when they were used in game theory and were applied to answer a problem in particle physics in the Manhattan Project, i.e., the development of the atomic bomb. It wasn't long before other scientists and engineers realized the power of Monte Carlo and began applying it many ways. But in the Hertz article Curran saw that for the first time someone was suggesting its use in business practice. This made so much sense to Curran that in 1968 he formed Decision Sciences Corporation which has pioneered the field of project risk analysis in the United States.

9.2   THE FOURTH VARIABLE

Curran observed that decision makers generally use only three building blocks on which to base their decisions. One is *units*, i.e., units of service required or rendered, units constructed, square meters of building space, etc. Another building block is *currency*, in whatever way you count it—euros, dollars, yen, etc. The third is *time*, time to complete a project, time to construct a building, etc.

Decision makers however often do not adequately consider a fourth variable, *risk*. What is risk? Here is the AACE International definition of risk (see Appendix B):

> Risk: The degree of dispersion of variability around the expected or "best" value which is estimated to exist for the economic variable in question, e.g., a quantitative measure of the upper and lower limits that are considered reasonable for the factor being estimated.

While technically correct, this definition does not explain what decision makers really need to understand. For decision makers it is better to define risk with an example—an example in plain English.

> Risk (Plain English Explanation): I think it will cost $1 million, but there's a chance that it might be a little higher or a little lower.

This is what decision makers want to know:

- What is the probable cost?
- What is the probability of the cost exceeding the estimate?
- If it does exceed the estimate, how high can it go, i.e., what is the economic exposure if the project overruns?
- If the cost underruns the estimate, how low can it go?

Answering those questions is what risk analysis is all about. Even if a project looks potentially very favorable, if the probability of the project succeeding is low or if the exposure is excessive, few managers will be willing to authorize the project.

When managers consider the first three of these variables—units, currency, time—they are dealing in the world of accounting. There is no uncertainty in accounting. It is precise. In accounting, if all of the rows and columns don't add up, you work day and night to correct things so that they will. You remove the uncertainty. That can't be done in decision making. Uncertainty always exists. When you include probability in your discussions, it is no longer the world of accounting. It is the world of decision making and there is a vast difference.

9.3   THE MONTE CARLO APPROACH

To perform a risk analysis does not require extensive knowledge of statistics and probability theory. All that is needed is to understand two key issues:

1. The probability number can influence decisions
2. You can measure the variability of a quantity as a simple range. How low can the variable go and how high can it go?

Now when should a risk analysis be performed? Certainly it is not always appropriate but if there are multiple numerical uncertainties and these uncertainties cause concern, the use of risk analysis techniques is advisable.

The question often arises as to how a highly sophisticated technique such as Monte Carlo simulation can possibly apply to projects for which much information is not precisely defined in advance. After all, Monte Carlo simulation was developed for very precise applications (like the atomic bomb) and, as such, generally requires great accuracy and precision in data inputs and well-defined probability density functions for the various variables. By comparison, cost and schedule information on engineering projects is rarely precise. Estimates are opinions of probable cost, not highly accurate answers.

The reason Monte Carlo approaches can be applied to engineering projects is that decision makers don't expect precise answers. They generally are willing to accept variances of 5 to 10% when making decisions related to quantitative values involving risk. By comparison, in highly scientific fields, errors of this magnitude cannot be tolerated. Such is not the case in engineering and construction. It's a different world. Because it is a different world and because we are more tolerant of some error in engineering or construction projects, we can deescalate the requirements on the part of the user of Monte Carlo.

The risk analyst must understand that the primary concern in the minds of managers is how large the economic exposure is if the project is authorized. Few managers will accept a project with a low probability of success and a high exposure. Managers rarely are risk takers. They are risk averse.

Often the worst case scenario approach is used to estimate exposure. In this approach, all major variables are estimated at their most extreme unfavorable values and the exposure is calculated. The results are inevitably horrible. They are the theoretical worst case. They are so mathematically remote that at best they are useless; at worst they are misleading. Even in a highly risky project, not every variable will go to its unfavorable extreme value. Some variables will show little variance from the plan and others may deviate in the favorable direction. Therefore the actual exposure generally is much less than the worst case scenario. Monte Carlo Simulation can identify what the actual exposure really is and what the probability of success is.

If, for example, you have performed a risk analysis and told management that they have a 80% chance of success but that they have a 45% exposure, what does that mean? That means that the bottom line decision variable can erode by as much as 45% of the target value.

"*Why?*" will be the next question they'll ask. "*Why will it be that way?*" They will want to look at a ranked list of risks and opportunities so that they can search for controllables and can challenge the management team to come up with alternative strategies and tactics. These alternatives can then be tested in the Monte Carlo environment to arrive at an optimum solution.

Monte Carlo is not only a great tool for evaluating a current plan; it's a marvelous tool for evolving a better one.

9.4   CONTINGENCY

In any estimate or project plan the estimators always include an item at the end of the estimate for contingencies. A contingency allowance is necessary because uncertainty exists in the estimating data and assumptions. The costs cannot be defined precisely when the estimates are made. To account for this, many companies tend to use an allowance of about 10% for contingencies in their estimates. In some cases, the 10% allowance is company policy. All their estimates include a 10% contingency. *This is totally fallacious.* The amount of contingency required is a function of the desired probability that the final project cost will not overrun the estimate. The higher the desired probability, the higher the contingency, and 10% is rarely the correct number. There is no logical reason to arbitrarily assume 10%.

With Monte Carlo analysis you are able to determine the correct amount of contingency. The amount of contingency needs to be balanced with the concept of confidence, i.e., confidence that you won't overrun your budget. If you want to have a lot of confidence, the contingency required will increase, and Monte Carlo analysis will define how much of an increase is required.

As mentioned earlier, an in-depth knowledge of statistics is not necessary in order to conduct risk analyses and range estimates using Monte Carlo techniques. There are a variety of PC-based software programs on the market for this purpose. Technically they are excellent, easy to apply, and answer all the key questions in the minds of decision makers. That is, they answer the key questions *if* the analyst uses the software properly. That is a big IF because software documentation is not always adequate to explain how to apply it properly to the imprecise nature of project estimating information. If care is not taken and the software is not applied properly, the result can be understatement of the true exposure in the decision. In such cases, actual risk can be far greater than predicted. In effect, the risk analyst actually can induce risk.

9.5 AVOIDING RISK ANALYST INDUCED RISK

The key to performing a project risk analysis is in properly identifying those variables that can have a critical effect on the project outcome and in applying ranges to those variables, and only to those variables. It is human nature to assume, for example, that a very large item in a cost estimate is critical just because of its magnitude. That is not the case. An item is critical only if it could potentially change by enough to have a significant effect on the bottom line. The effect need not be negative. What matters is if the effect is significant and can cause a change in either the negative or positive direction.

Curran (1988) has demonstrated that in virtually all project estimates the uncertainty is concentrated in a select number of elements—typically fewer than 30. In most cases there are only 10 to 20 critical variables, even though the estimate may have hundreds or thousands of variables. You must understand which variables are the critical ones. Unless you take the time and trouble to measure the uncertainty of each of them, and only them, and unless you take the time and trouble to combine those uncertainties into some bottom line uncertainty using a technique like Monte Carlo, you will never get to the right answer.

Not everything is important. Very few things are really important. This is called variously the *Law of the Significant Few and the Insignificant Many* or the *80/20 Rule*. Others refer to it as *Pareto's Law* after the noted Italian sociologist and economist Vilfredo Pareto. Long ago Pareto examined how wealth was distributed from country to country, concluding that in any given country a small percentage of the population would collectively account for most of the wealth. The fact of the matter is that this principle applies well beyond economics. It extends to almost all issues involving multiple variables in any area of investigation.

How then do you identify which variables are the critical ones? Quoting Curran (1988):

> . . . a critical element is one whose actual value can vary from its target, either favorably or unfavorably, by such a magnitude that the bottom line cost of the project would change by a amount greater that the critical variance . . .

Curran goes on to define critical variance empirically as "in the neighborhood of 0.5% in conceptual estimates and 0.2% in detailed estimates." He goes on to state that "If the bottom line measures profit rather than cost, the threshold values are approximately 5% and 2% respectively." Thus, for example, if a detailed estimate is $1,000,000, the critical variance is $2000. Any variable which could cause the bottom line to change, up or down, by $2000 or more would be defined as a critical variable.

It is very important to understand that the magnitude of a variable is not important. What is important is the effect of a change in the variable on the bottom line. Relatively small variables are often critical while very large variables may not be critical at all. Typically there will be only 10 to 20 critical variables, even in the largest projects with hundreds of variables to consider. In identifying the critical variables it is also necessary to link strongly related variables together, not to treat them separately. It is also necessary in the range estimate to apply ranges only to the variables that are identified as being critical. You must know when a variable is important and when it is not.

9.6 CONDUCTING THE RISK ANALYSIS

Now, what are we going to do with the items that have been identified as being critical? Give them a range and make an estimate of the probability that each item can be accomplished within the original plan, i.e., for a cost item, the probability that it can be accomplished at a cost no greater than the originally estimated amount.

And how do we do that? The very best way is get everybody involved. Everyone who understands the critical variables should be there contributing to the process except for anyone who may have undue influence on the group. Don't include them because it will taint the results.

The process is not as time-consuming as it may appear. In many cases, good risk analyses can be performed in a matter of hours. If the process takes a long time, there is probably something going wrong. That's not to say there aren't cases where it will require a lengthy period of time to do a risk analysis. But in the normal course of events, that doesn't occur. This should be a decision based upon the collective experience of the group of people who understand the variables.

If you only ask a very inexperienced manager to range a critical element, you inevitably will get a very big range. Lack of experience breeds excessive conservatism.

On the other hand, if you only ask the estimator or quantity surveyor who came up with the figure, you are going to get a narrow range. It is human nature to defend your estimate. To agree to a wide range is akin to admitting failure.

If you recognize these human weaknesses and structure the ranges properly using the collective experience of the group, you will learn:

- The probability of having a cost overrun
- How large the overrun can be (the exposure)
- What can be done to capitalize on opportunities and reduce risk

Most importantly, you will learn how much contingency to add to the estimate to reduce the residual risk to an acceptable level.

9.7 AVAILABLE RISK ANALYSIS SOFTWARE

There are some excellent pieces of software for performing risk analysis. Three of the better known ones are REP/PC, @RISK, and Crystal Ball®. These three software packages are listed as examples only. There are many others that may be suitable for any particular application.

Here are some of the characteristics of the three programs:

- REP/PC, Decision Sciences Corp., St. Louis, MO—standalone software. Input: ranges and probabilities.
- @RISK, Palisade Corp., Newfield, NY—spreadsheet-based software. Input: probability density functions.
- CRYSTAL BALL®, Decisioneering® Corp., Denver, CO—spreadsheet based software. Input: Probability density functions.

Of these, REP/PC is preferred by many companies for engineering and construction work. However, all three packages are excellent if applied properly, but there are limitations to be considered with all of them. Failure to properly identify the critical variables, or to assume that some variables are critical when they are not, will yield an incorrect analysis and will understate risk. REP/PC has proprietary algorithms in the software to detect improperly identified critical variables. The other two pieces of software do not protect you against this problem. You must follow the rules outlined earlier for identifying the critical variables. The software itself will not do this job for you.

REP/PC requires the ranges and probabilities for the critical variables as the input. This program unfortunately is a DOS program and is not available in a Windows® version. That is not a serious limitation for versions of Windows up through Windows Me, which directly support DOS. Unfortunately, the program is not available in a version which is compatible with later versions of Windows®.

@RISK and CRYSTAL BALL® are fully Windows compatible and link directly to Excel or other spreadsheet software but do require that probability density functions be identified for each critical variable. This is generally not a problem however, because we are not dealing with rocket science. High precision is not required for the typical projects cost engineers and project managers work on. If the probability density functions are undefined (as will usually be the case), a simple triangular distribution can generally be safely assumed for each critical variable. This assumption is sufficiently accurate for cost work.

Risk analysis is not difficult to perform properly, and the benefits in project planning are great if the analysis is properly done. However, if it is not done properly, the results can be disastrous as the analysis can severely understate risk and lead to unsatisfactory conclusions about project viability.

REFERENCES

Curran, M. W. (1976). Range estimating: coping with uncertainty, paper I-6. *AACE Transactions*. Morgantown, WV: AACE International.

Curran, M. W. (1988). Range estimating: reasoning with risk, paper N-3. *AACE Transactions*. Morgantown, WV: AACE International.

Curran, M. W. (1988). Range estimating: contingencies with confidence, paper B-7. *AACE Transactions*. Morgantown, WV: AACE International.

Curran, M. W., ed. (1998). *AACE International's Professional Practice Guide to Risk*. Vol. 3. Morgantown, WV: AACE International.

RECOMMENDED READING

Bent, J. A. (1996). Humphreys, K. K., ed. *Effective Project Management Through Applied Cost and Schedule Control*. New York: Marcel Dekker.

Schuler, J. (2001). *Risk and Decision Analysis in Projects*. 2nd ed. Newton Square, PA: Project Management Institute.

10

Cost Engineering in a Worldwide Economy

10.1 INTRODUCTION

We can no longer consider anything we do in business as a purely national thing, constrained by the borders of our own country and by a single currency. Everything we do must be considered in light of international economics.

Industrial firms are increasingly multinational, and many companies now build and operate plants in several nations. Economic cooperation agreements between nations, such as the European Economic Community (EEC), the North American Free Trade Agreement (NAFTA), and the General Agreement on Tariffs and Trade (GATT) are encouraging further industrial globalization. Many European countries have even abandoned their national currencies in favor of a common currency, the euro. The franc, lira, guilder, Deutsche mark, and many other currencies are no more.

It is becoming increasingly important for project control professionals to be familiar with techniques for estimating costs in other countries and to be able to compare costs in different nations. The questions that are always being asked are: "What will this plant cost in the home country?" "What about Canada, Great Britain, South Africa, and Australia?" and "Which location is the most attractive financially?" To this end, international location factors, such as those which are discussed in Chapter 7, are useful but they do not go far enough. They only tell the cost professional the approximate cost to replicate a facility in one country

---

This chapter is based upon a paper by Charles P. Woodward and Kenneth K. Humphreys, The Planning and Execution of an International Project: A Checklist of Actions, which was prepared for presentation at the 15th International Cost Engineering Congress, Rotterdam, 1998. Mr. Woodward was the primary author of the paper. His current affiliation is as Cost and Project Controls Engineer, Burns & Roe Services, Inc., Virginia Beach, Virginia.

given the known cost of an identical facility as built in another country. Location factors cannot distinguish between different designs, construction conditions, climates, geological conditions, etc. Considerably more information must be obtained if the cost engineer is to produce an accurate estimate for an international project. The planning and execution of an international project requires many special considerations not usually encountered when planning within your own country. There are the obvious differences such as wage rates, productivity, duties, and taxes, but many more differences may be overlooked. To properly prepare an international estimate and to avoid unpleasant surprises requires that you conduct research on construction practices in the target country. This chapter presents many of the differences that need to be considered and recommends actions which should be taken in order to insure development of a reliable and accurate estimate.

10.2 PROJECT ELEMENTS CHECKLIST

10.2.1 Project Design

1. What local codes and practices apply to the project design? For example, do you construct prefabricated steel buildings or concrete block buildings?

2. What weather and climatic conditions are likely to prevail at the project site? How do these differ from your home country? Is the project location subject to temperature extremes, severe rainfall, flooding, etc.? (e.g., monsoons in Southeast Asia; frost levels in the north; permafrost in the far north)

3. Must consideration be given to any special geological conditions such as earthquake zones, unusual soil conditions, etc?

4. What will be the language of the contract documents, drawings, etc.?

10.2.2 Bulk Materials

1. What materials are available locally? Is the quality adequate? For example, in some sandy desert areas such as Saudi Arabia local sand may not be suitable for use in cement and concrete. Sand may have to be imported. It may sound like sending coal to Newcastle but it isn't—it can be a major concern.

2. Is it cheaper to buy locally or import? Before the devaluation of the Mexican peso, there were situations when it was actually less expensive to import reinforcing steel into Mexico than to purchase it there.
In some situations there may be a VAT on local materials while the project is exempt from duties. This may make it cheaper to import than buy locally. Often owners can take title to goods at the point of import and avoid duties and taxes. Also, owners can sometimes recover the VAT.

3. Should you use different practices? For instance, if construction labor costs are very low, it may be more economical to fabricate rebar at the site

rather than at the factory. Similarly, in some areas hand mixing and placing of concrete may be more economical than using ready-mixed concrete.

10.2.3 Engineered (Process) Equipment

1. While equipment pricing tends to be international in nature, there are places in the world that are more competitive than others. For example, for one project, a gas turbine generator was priced at U.S. $14,000,000 if purchased in the U.S. but if ordered in India it could be bought from the same U.S. company for U.S. $12,000,000 because of competition.

If possible, obtain quotes for your specific location rather than using recent pricing in other geographic areas. If information from other areas must be used or if importing of equipment is required, be sure to factor in shipping costs and duties from each potential supplier location.

2. As indicated above, be sure to evaluate the total cost in selecting your equipment supplier. In addition to the base price, you should consider shipping costs, duties and exchange rates.

3. Be sure to evaluate your spare parts requirements. Due to long delivery times for certain components, particularly those that must be imported, you may want to stock more spares. For example, one U.S-based company maintains a spares inventory of about U.S. $500,000 for its domestic gas turbine generator projects. For projects in other countries, this same company allows U.S. $2,000,000 for spares.

10.2.4 Construction Labor

1. Is there skilled labor available? Does the labor force have the skills needed for the project? Skilled trades people may be available but technically trained people may not be available.

2. What is the local productivity? While there are several publications that quote average productivity compared to a base location such as the Houston Gulf Coast area of the United States, wide variances may exist in actual productivity in any given location. For example, in Saudi Arabia, the average labor productivity factor is about 1.6 versus the U.S. but may vary from being similar to that of the U.S. to over 3.0 depending upon project conditions and the mix of local and expatriate labor. Within the U.S., productivity also varies widely.

3. Will you be required to provide housing and services, such as medical facilities, as a part of your construction program? In some areas, meals and shower facilities are required. In Brazil, for example, breakfast, lunch, and transportation to the job site are usually required.

4. How do you obtain local labor? You may be required to pay fees to a labor broker for staffing your project. Subcontracting may be preferable.

5. What are local payroll taxes? They can exceed 100% of payroll in some countries.

6. Should you import construction labor? Is it legal to import labor?

10.2.5 Construction Equipment

1. Is construction equipment available locally? If so, what is the cost? If equipment is in short supply, the costs may be much higher than in your home country.

2. If you need to bring in equipment, will you be required to pay duties? Sometimes duties can be avoided by posting a bond to guarantee that you will export the equipment when the job is complete.

3. What are the rules for removing equipment after the project? Sometimes exporting will be prohibited—this is a big potential risk.

4. Are the local construction forces trained in the use of the equipment? In some areas workers may not know how to use power tools. Hand tools may be required.

5. Supposedly trained local workers may not have adequate skills. On a recent Chinese project, over 70% of people holding Chinese certificates of qualifications were unable to do the work for which they were hired.

10.2.6 Construction Management Staffing

1. What positions will be filled with expatriates versus local hires? Use of expatriates is expensive. If expatriates are hired, they should be bilingual. Seek out bilingual local hires for trial as possible permanent employees.

2. What living arrangements will be provided (camp versus local housing) and what is the cost? Will food be a per diem expense or will workers be fed in a galley? What about security? Do you need a gated camp? Alarms? Home systems?

3. What special services, such as housekeepers, drivers, etc., are required?

10.2.7 Schedule

1. How long will project mobilization take? Project mobilization may take longer than normal if you are setting up a base camp, bringing in construction equipment, etc.

2. What is the schedule effect of equipment delivery time? Allow for additional delivery time for imported equipment to cover ocean freight, offloading, customs clearance and possible delivery to the site. A self-unloading ship may be required if port facilities are not adequate.

3. Will weather introduce any special schedule considerations? Be aware of local weather conditions, such as monsoon periods in the tropics, that may delay the schedule.

4. How will local productivity, work practices (e.g., carnival), culture, religion (e.g., prayer breaks), and workforce level of education affect schedules?

10.2.8 Local Infrastructure Requirements

1. Are local water, sewer, electricity and other services available? If not, will on-site facilities (e.g., generators) be required? In China, for example,

local water is generally available but it may contain sediment and may not be potable.

   2. Are highways and rail service available? Are they adequate or do they require upgrades? Roads and bridges, for example, may not be able to carry large loads.

   3. Are there relocation requirements for people living on the site? What are these relocation requirements? Often people must be relocated and provided with new housing. Consider the Three Gorges Dam project in China which displaced 1.13 million people.

   4. Can the local port handle the cargo? Is the depth adequate? Is the dock adequate? Can the cranes handle your loads? Will self-unloading ships or crane barges be required?

10.2.9   Material & Equipment Delivery (Shipping & Customs)

   1. Should shipments be consolidated? Consolidation of shipments can save significant shipping cost but can impact your schedule.

   2. What fees are required for customs clearance? They can be significant. As an example, according to Aspen Richardson's *International Construction Cost Factor Location Manual* (see Appendix C), the following is a list of duties and fees that may be charged for imports into Brazil (as of January, 2003):

> Duties 0–40%; average 14%
> Merchant Marine Renewal Tax 25% of freight charge
> Syndicate fee 2.2% of c.i.f.*
> Brokerage fee 1% of c.i.f.
> Warehouse tax 1% of import duty
> Fee for handling charges $20–$100
> Administration commission $50
> Import license fee $100
> Additional port tax 2% of c.i.f.

   As can be seen from this list, the costs of importing can be very significant, but exceptions are often available and need to be explored. India, for example, levies duties of 40 to 100 percent or more, yet has granted concessional rates as low as 20 percent for recent power projects.

   Some items may be restricted for import requiring use of locally manufactured items. For example, in China switchgear and fire protection equipment must be Chinese made.

   3. What is the cost of transit insurance to move the equipment to the site?

   4. How will you get the equipment from the port to the site (rail, highway, barge)? If you are moving the equipment by highway, can the roads and bridges

---

*Price of goods plus packing, freight, insurance, and seller's commission.

handle the loads or do they need reinforcement or bypass? Is there special equipment available to handle the heavy loads (off the ship and on the highway)?

5. Do freighter schedules meet your project schedule? Heavy lift ships have limited availability and, if required, may present schedule problems.

6. What security is required to ensure that your equipment arrives at the site? Is theft a potential problem? Must equipment be guarded?

7. Do you require special shipping due to project financing? For example, U.S. export–import bank financing may require use of U.S. flag ships at approximately a tripled freight cost.

10.2.10   Progress Payments (Cash Flow)

1. What up front payments will be required? Contractors in developing nations may be cash poor and may require a larger than normal up front payment to help with project operating capital. A 15 to 20% advance payment is not unusual. This may be necessary to enable them to buy materials. Funds may not be available locally to them and, if they are, the loan interest rates may be very high.

2. If you are being paid by a foreign government, what kind of processing time can you assume for invoice payments? The bureaucracy can be very slow.

3. What currency will you be paid in? If you are paid in local currency, there will be a conversion fee and, on occasion, convertibility may not be available.

4. If payment is in local currency, what risk exists for significant fluctuation in exchange rates or for changes in government policy which might affect exchange rates? Consider the four Asian countries (Thailand, Phillipines, Malaysia, and Indonesia), for example, that suffered losses of 30 percent or more in the value of their currencies in three months during 1997. All four currencies had a long history of stability before this sudden decline.

A good contract clause is one that specifies bids are to be adjusted for currency reevaluations, for example, the change from pesos to new pesos in Mexico.

5. What is the schedule of payments to local workers: daily, weekly, monthly?

10.2.11   Local Taxes

1. What local business taxes must you pay for working in the country? Some countries levy substantial taxes in order to remove profits from the country.

2. What local sales, property, VAT or other taxes will be assessed against your project? VAT is a tax on a tax. For example, the contractor buys materials and equipment and pays the VAT. The owner is billed and pays the VAT again. Sometimes the VAT can be refunded but it is best to avoid it if possible by the owner buying the materials directly (or by taking title at the time of import).

10.2.12   Insurance

1. Does the local country allow for private insurance? If not, you may need to contract with government agencies in the host country for insurance rather than with your regular carrier.

2. What special insurance is required for international freight losses? Marine freight insurance may be included to the site but what about project delays if equipment is lost? Insure for delays plus emergency replacement costs plus interest during construction delays.

3. Is bonding available in the country? If not, a letter of credit may be required.

10.2.13   Project Finance

1. Does project financing require that a certain percentage of the project be purchased in certain countries? U.S. export–import financing may impose such requirements.

2. On joint ventures, get a majority interest. Otherwise local codes and standards will apply, and they are not necessarily of the level desired. On one Chinese project, building to Western design standards cost 5000 RMB/m^2. If built to Chinese standards, it would have been 1500 RMB/m^2.

10.2.14   Legal Recourse

1. What recourse do you have if your customer in the country defaults on the contract through cancellation, nonpayment, bankruptcy, etc.?

2. What legal recourse do you have for nonperformance of local subcontractors? What about lawsuits?

3. What about permits? Often permission must be obtained from many government authorities before the project can commence. For example, in Suzhou, China, the Lion Nathan Brewery which opened in 1998 was required to obtain 41 different government construction permits.

10.2.15   Social System

1. What local holidays are observed?

2. Is family housing required at the job site? Indonesia, for example, encourages family status.

3. How do local religious customs impact the work schedule? For example, are prayer breaks customary?

4. What about local work customs such as long midday work breaks in some countries (e.g., Mexico and Latin America) or exercise periods (e.g., 1/2 hour per day in Japan)?

10.3   HOW DO YOU BEGIN?

If you have never tackled a foreign project before, the above list is certainly enough to discourage anyone from attempting it. Fortunately, there is a good support infrastructure to help you get started. The following are some recommendations to help you prepare for a foreign venture:

> Most major industrial nations have the equivalent of a Department of Commerce. These government agencies have staffs that are well versed in helping you develop a plan and obtain the information you need for a successful project. You should begin your project by contacting one of these agencies for assistance.
>
> You should consider a local partner or consultant in the host country. This person or company should have the ability to help you with the language, identify local rules and regulations that will apply to the project, and help you through the processes.
>
> For local construction practices, pricing, labor, etc. you will need help from someone skilled in local construction practices: a design firm, local contractor or just a special consultant. Discuss with this person the project logistics of local contractors vs. self perform, construction equipment, site housing requirements, local versus imported materials and labor, schedule, etc. to develop a project plan. To help locate this person or firm, you may wish to start with the International Cost Engineering Council (ICEC). ICEC has member societies doing business in much of the world and you may be able to obtain recommendations or assistance from the ICEC members. You can contact ICEC on the World Wide Web at http://www.icoste.org or through the cost engineering, quantity surveying, or project management society in your country. Table 7.9 provides a listing of ICEC members. Appendix C also provides a listing of web sites for the ICEC members and the members of the International Project Management Association (IPMA).
>
> Once you have put together your basic project plan, you will need to begin gathering the specific cost and schedule information required for the project. Here are some suggestions to help you compile the information:
>
>> For specific local pricing information, the best approach is to obtain bids from local contractors. Put together a bill of quantities and work through your local construction consultant to obtain the bids.
>>
>> If you are working on a technology project in a developing country, do not assume that the contractors you are working with understand the project concepts and how they go together. You may get some very strange results if you just supply a list of quantities and ask for unit pricing. Educating your project subcontractors can provide much better results.
>>
>> Before you visit your subcontractors, provide them with the quantity list and give them some time to review it. When you visit, take photographs of similar work, layouts and elevations of the

proposed project, and a simple project schedule. If you have time-lapse film of the construction of a similar facility, it will be of great value in educating the subcontractors.

When you visit the subcontractors, take time to sit with them and discuss the quantity list to make sure that they understand your terminology. Remember that an estimator who does not understand a project will be more conservative and will include greater contingencies in the estimate than may be justified.

For equipment pricing, be sure to remember to consider any specific country sourcing due to finance requirements. Look to multiple country sources for bids. Many times the equipment vendors can arrange special financing or in-country manufacturing to avoid duties or can provide other assistance to the project. Be specific that you will be evaluating the bids on a completed project cost basis and not merely on bid price.

Select a freight forwarding company early in the planning stage to help with the schedule logistics, freight costs, and duty requirements for the project. These companies generally earn their fees, as travel consultants do, by commissions they receive from the freight lines. You should be able to find a company that will work with you on that basis. You should select a company that has offices in the countries of origin for your equipment and in the country where you will be working. The local agents will be very valuable with helping you through the customs and local transportation requirements.

There are several international consulting companies that provide information about taxes and business requirements around the world. Work through your regular accounting firm to identify a company to work with. You will need to obtain reports on local taxes and business issues for your project.

You will need to obtain an international attorney to help prepare the contracts, business plans, and other legal documents for you to do business in your host country. It is important to find an attorney who is experienced in international law and preferably one who has done business in your host country before.

10.4 CONCLUSION

The best possible summary for this discussion is from John R. Barry's paper, "Ten Commandments of International Cost Engineering," presented at the 1993 Annual Meeting of AACE International:

The reliability of available data is usually suspect, so test, test, and retest it. Communication problems are difficult across cultures, and sometimes language differences can cause misunderstandings when soliciting data.

The use of international factors is a session all its own. A few basic principles for guidance include keeping in mind that a time or place factor means nothing without an exchange rate and date and that a base city must be identifiable, since variation can exist within your home country.

The paper concluded with Barry's "Ten Commandments for Worldwide Cost Engineering," which have appeared in several of his articles over the years. They are a fitting conclusion to this discussion as well.

John Barry's Ten Commandments for Worldwide Cost Engineering

1. Thou shalt not begin an international cost engineering assignment without preparing for the differences in culture and protocol.
2. Thou shalt not ignore investment objectives of the target country's government.
3. Thou shalt not look at building your own facilities as the only way to enter business in a country.
4. Thou shall not use biased estimated scope that does not reflect technical, cultural, legal, and climate differences.
5. Thou shalt not ignore what equipment must be imported and the impact on cost and schedule.
6. Thou shalt not accept as gospel cost data for other countries without thoroughly checking for understanding and testing for reasonableness.
7. Thou shalt not ignore productivity, weather, religious practices, and construction methods when calculating labor cost.
8. Thou shalt not ignore the additional risks associated with cost and schedule on international projects.
9. Thou shalt not forget that AACE and ICEC members are valuable resources who are capable and willing to help.
10. Thou shalt not ignore the previous nine commandments.

International construction carries with it many risks, but as economies become more global, the reward of contracting internationally will become greater. The above list of activities and recommendations is far from complete, but it does cover most of the more significant items of concern. If you begin working with the above recommendations, you will have the tools needed to make you successful in international project work.

RECOMMENDED READING

AACE (1998). Symposium INT—International Projects. *AACE International Transactions*. Morgantown, WV: AACE International.

Barry, J. E. (1993). Ten commandments for international cost engineering. *AACE International Transactions*. Morgantown, WV: AACE International, Paper L.4.

Humphreys, K. K., Hamilton, A. C. (1999). The international cost engineering council: a worldwide source of cost and project information. *AACE International Transactions*. Morgantown, WV: AACE International, Paper INT.01.

11

Computer Applications for Project Control

11.1 WHY CONSIDER COMPUTERIZATION?

Computers have become such an accepted part of our everyday lives that it becomes easy to lose perspective on the fundamental issues of computer use: why they are or are not appropriate in a given situation, what they should and shouldn't be expected to do, and how to achieve the computer capability needed. These questions have become markedly more complex in recent years as the computer industry has offered many new possibilities while simultaneously raising new questions.

Fortunately, in spite of the many new developments in computing, the fundamentals of system design, development, and implementation have not really changed. Equipped with these fundamentals, we can use the latest computer technology in an effective way, regardless of how rapidly or dramatically the technology changes.

This chapter covers the fundamental concepts of system design, development, and implementation as they apply to project control. Like other chapters in this book, this one is not intended to make the reader an expert. Rather, it is intended to give cost and project engineers or managers all they need to know to be able to assure that effective computer-assisted project management and control techniques are available for each project. For that reason, we deliberately retain the perspective of the computer system *user* throughout this chapter.

Our first consideration as users should therefore be: "Why consider computerization in the first place?" If we look at the most effective computer systems

This chapter is adapted from Chapter 14 of the book Computerized Management of Multiple Small Projects, by Richard E. Westney, Marcel Dekker, Inc., New York, 1992.

and consider the inherent capabilities offered by computers, we find some or all of the following general conditions for effective information systems:

- A considerable amount of data that has to be handled
- A requirement to file data and present selected data when needed
- A short time available to process data
- Relatively simple and repetitive operations for handling data
- A large number of people in the organization involved with the data going into the system, the information presented by the system, or both
- The consideration that the availability of timely and accurate information is essential to the achievement of business goals

It is certainly evident from the above that project control and project management fit all the criteria for effective computerization. We have a large volume of data to handle in a short period of time, a large number of people involved, a need to be able to store data for later use, a lot of data needing manipulation using simple and repetitive calculations and, most important, a need for timely and accurate information. Computer power can be a time saver in:

- Network analysis and scheduling
- Resource analysis
- Cost estimating and expenditure forecasting
- Management reporting
- Decision making

Therefore, computer-assisted techniques deserve close consideration in the project environment.

11.2   DEFINITION OF TERMS

To avoid confusion or misunderstanding it is worthwhile to define the terms we will be using.

Project management system: A computer system for processing project data in order to provide information to facilitate project management actions.

Project management method: A series of specific actions to accomplish a project management task. The method may include use of a project management system and is documented in a uniformly followed procedure on each project.

Data: Numbers and words that represent some aspect of the project. "Raw" data is in the form in which it was captured; "processed" data has been subjected to some manipulation.

Information: Processed data presented in a way that conveys specific knowledge that will enable the person receiving it to perform project management functions. Information can be presented in printed reports, with graphics, or on a screen display.

Input: The entry of raw data into the system. Input may be by means of a keyboard or electronically through a communication network.

Output: The presentation of information prepared by the system. Output can be in the form of printed reports, management graphics, a screen display, an electronic message to another computer, and/or a file on a disk or tape.

Random Access Memory (RAM):   The part of a computer in which data is processed. The amount of RAM indicates the size of the program and the amount of data that can be handled in memory. The contents of RAM are temporary.

Read Only Memory (ROM):   That part of the memory containing instructions to the computer. It cannot normally be accessed or changed by the user. The contents of ROM are permanent.

Bit: A measure of the quantity of data stored or processed. One bit is one yes/no signal.

Byte: A measure of the quantity of data stored or processed. One byte equals a word consisting of eight bits and is sufficient to describe a character or number. A *word* is the number of characters that the computer can process together.

Hardware: The combination of copper, steel, plastic, and silicon that makes up the physical part of a computer. Hardware alone cannot perform any functions.

Operating system: The programming built into a computer that enables it to function. The operating system tells the hardware what to do and how to process data with its memory, disks, keyboard, and other devices. The operating system also tells the computer how to interface with the various items of software which will be run on it. It is important because it defines the type of software that can be used, as well as the type of programming that can be done.

Program: A set of instructions for the computer to follow. A program defines the input, processing, and output that must be performed. Programs are written in various computer languages in which the instructions are expressed in a way that can be interpreted by the operating system.

Software: A package of one or more programs designed for a specific application. Software will work with one or more operating systems and hardware configurations. General application software is sold as a stand-alone product for general use. Specific-application software may be purchased, written as a new program, or, in some cases, adapted as a combination of both from shelf packages.

Bundled system: A computer system in which hardware and software are sold together in one integrated package. In a bundled system the hardware and software cannot be separated.

Unbundled system: A computer system in which hardware and software can be purchased separately. The software is said to be "machine portable" in that it will usually run on more than one type of hardware. One advantage of an unbundled system is that it enables the purchaser of the hardware-software package to use the hardware for other applications or to purchase only the software.

Local Area Network (LAN):   A means of linking together several computing components, such as a large mainframe computer, several terminals, several personal computers, printers, and plotters. The purpose of an LAN is usually to allow users to communicate, sharing specialized equipment (such as a plotter) and share data. Networking can be accomplished with special hardwiring, use of telephone lines, satellite communication, or special networking packages, including both software and hardware.

Terminals: Local workstations at which the system user can input data using a keyboard and receive output in the form of a visual display and/or in print. A "smart" terminal (usually a personal computer) is also capable of storing and processing data itself. It can therefore upload data that it has stored and/or processed to another computer or to other terminals, as well as download data which it receives from another computer. A "dumb" terminal is simply an input/output device and has no internal storage or processing capability.

Mainframe: A large, powerful, centralized computing facility supporting most if not all of an organization's data processing needs. It is characterized by high-speed processing, extensive storage capability, multiple input and output facilities, a wide range of software, complexity of operation far beyond the skills of the average user, rigid environmental requirements, and very high cost. The organization with a mainframe computer will inevitably have a major organization function (data processing, management information systems, etc.) that is responsible for servicing the various users. A mainframe is distinguished by its purpose, which is to provide massive calculation and storage capabilities to an entire organization.

Minicomputer: A powerful computing facility that can support all, some, or perhaps only one of the organization's data processing needs. The minicomputer is still a sophisticated system, but it is most apt to be in the hands of the user rather than in a data processing group. Most "minis" can function in any reasonable environment, are portable, and are far less expensive than a mainframe. A minicomputer's (or workstation's) purpose is to support a small group of users with reasonably powerful computing capability.

Personal computer (PC):   A small computer intended for one user. Because the PC is entirely in the hands of its user, it offers the benefits of convenience and responsiveness that larger, more complex systems cannot match. The very low cost of the PC not only makes it easy to implement but it also means that the risks of developing a system are small; the worst that can happen is that we scrap it and try again. It is because of these features of low cost, user convenience, and ease of implementation that PCs are a good candidate for small project applications.

Peripheral equipment: Devices that are external to the processing portion of the computer and provide input of data, output of information, storage, and special functions such as graphics.

Interactive system: A computer system in which the user interacts in real time with the computer. This interaction usually occurs with a terminal in which commands or questions are processed quickly and a result displayed. In comparison, the *batch processing* takes all the input at once and later delivers the printed output.

User-friendly system: A system that enables the user to interact easily with the computer with simple commands. User-friendly systems typically feature English-like commands instead of the usual coded commands, on-line tutorials in which the system instructs the user as he or she goes, and helpful diagnostic messages when the user makes an input error. User-friendliness is extremely important in the small-project environment since most users are not skilled computer operators. The *graphical user interface* (GUI) in which a *mouse*, a small hand-held device used to move a cursor on the screen, is used to issue commands and interface with images, provides a high degree of user-friendliness.

Menu-driven system: An interactive system in which the user is presented with a series of options from which to select, and then follows the instructions (or "prompts") as displayed on the screen. In effect the computer is telling the user what to do. Menu-driven systems are relatively easy to learn and use and are appropriate for users with little system training. The disadvantage of the menu-driven system is that it is relatively inflexible, a characteristic that can frustrate the skilled user and prevent full utilization of the system's capabilities.

Command-driven system: An interactive system in which the user has a variety of commands that he or she can employ to tell the computer what to do. This type of system is more flexible than a menu-driven type and enables the user to obtain exactly the results needed. However it also requires more sophisticated computer knowledge on the user's part. Many systems are available that have both menu-driven and command-driven capability; this can be very useful in the small project environment. For example, data from timesheets could be inputted by a clerical person with little training using a menu-driven routine, while decision analysis and cost and schedule forecasting could be done by a project engineer using a command-driven routine. Often a set of commands internal to the software, a *macro language*, can be used to create menu-driven repetitive tasks.

11.3   CONSIDERATIONS IN THE DESIGN OF PROJECT MANAGEMENT SYSTEMS

In this discussion of project management systems, remember that a computer cannot manage a project. What a computer can do is make the people responsible for the project as effective as possible. The essence of project control boils down to the timely processing and presentation of useful information that can be summarized as follows.

A project management system is a means of presenting (1) the right information, in the (2) right format, to the (3) right person, at the (4) right time.

The cost engineering/project management or project control functions that benefit from computer assistance include:

In the project evaluation phase:

- Cost estimating and expenditure forecasting
- Planning and scheduling
- Resource scheduling
- Multiple-project analysis and scheduling
- Economic analysis
- Bid preparation and analysis
- Management-level decision-making
- Collection of project data
- Design and technical analysis
- Document control
- Contractor and bid evaluation

In the project-execution phase:

- Cost control
- Progress measurement and schedule control
- Productivity analysis and control
- Project-level decision making
- Materials management
- Control of labor and equipment resources
- Cost and schedule forecasting
- Contract administration
- Design decision making
- Document control
- Quality control
- Risk management
- Field supervision
- Management reporting

To build a project management system for a specific application, all we really have to do is combine the basic elements of system design to suit our requirements. These elements are:

1. *Input* in the form of keyboard entry, data from a storage disk or communication with another computer
2. *Processing* or data manipulation using equations and a programmed procedure
3. *Storage* of data or results on a tape or disk
4. *Output* in the form of displays, reports, and/or graphics

As cost managers all we really need to know about system design is how to define and combine these four elements of a system such that the computer does exactly what we want it to do. Our system design becomes a specification that will be satisfied either by purchased software or by a program that we will have written or by a combination of the two. The software will then be installed in some hardware that always fits our specified requirements.

However, before a system can be designed, it is necessary to address some very fundamental questions about the work we do involving how we do it now, how we should do it, and how we could do it with computer assistance. In other words the essence of system design is not programming, nor is it selecting the hardware and software. It is defining the job the system is to do. To develop that definition, we have to understand our own jobs very well.

11.4   BASIC STEPS IN SYSTEM DESIGN AND IMPLEMENTATION

To those of us involved in project management and cost control it may seem like a lot of unnecessary work to follow all the system design steps listed below. After all, all the system vendors claim their product will solve all our cost engineering/ project management problems, so why not just go ahead and pick one? The number and diversity of available computer systems is, in fact, one reason why the preparation of a system's specification is so important. The only way to cope with the volume of information, the pressure of time, and the intensity of salespeople is to have a clear idea of exactly what we want.

The system's design is also important because projects exist within an organization, and the power of the organization is mightier than any computer. Therefore, if our system is to be effective, and if it is to make our hard-won project management methods effective, it must fit the organization and enhance its operation. To assure that this is the case, we must know ourselves, our jobs, and our needs—as well as those of the organization in which we work—before we can know what computer system will work for us.

The basic steps in system design are as follows:

1. Define user requirements.
2. Develop system performance specifications.
3. Prepare plan and cost estimate for implementation.
4. Obtain management approval to proceed.

5. Prepare sample project and reports.
6. Contact suppliers and develop a short list.
7. Conduct a "compute-off" and evaluate results.
8. Obtain management approval for implementation.
9. Implement system.

Each of these steps is described in detail below.

Step 1: Defining User Requirements

This first step is both the most important and the most difficult. It requires a thorough analysis of current and future job practices as they relate to the receipt and manipulation of data and the presentation of information. The "primary users" referred to below are the typical people in the organization function for whom the system is designed (e.g., the project engineer). Other people in other functions might also be users; we will refer to them as "secondary users." Taking the perspective of the primary user, we can develop the necessary definitions by considering the following questions.

Reviewing and Defining Existing Practice

- Who will be the primary users of the system?
- What are the objectives of the primary users' job functions? What methods are now used to perform these functions?
- What information or data do the primary users get now? From whom and in what form do they get it?
- What information do the primary users provide to others? What information is provided to whom? To what use is it put?
- What data manipulation is now performed by the primary users in order to provide the necessary information to others?
- What problems exist with current practice?
- What improvements could be made to the current practice?
- Who would the secondary users be? What enhancements to current practice will meet their information needs? What information must they provide to others?

Defining Ideal Practice

- Can or should the primary user's job function be changed? If so, what should it be?
- What methods should ideally be used?
- What information do the primary users need to do their jobs?
- To what organizational functions should the primary users provide information? In what form should it be provided? How would it be used?
- What data manipulation should the primary users perform to provide this information?

- Who would be the secondary users in the ideal situation? What information would they receive? What would or should they provide to others?
- What improvements would the ideal practice achieve over current practice?

Specifying Short-range System Requirements

- What existing user functions could be computerized for greater efficiency?
- What immediate improvements could be gained by computerization?
- What existing computer systems might be used?
- With what existing computer systems might the new system interface? What would be the type of data exchanged across the interface?
- What level of computer capability will the primary users have? What other levels of user capability need to be accommodated?
- What organizational functions will be putting in what data?
- How much data needs to be stored? How much of that amount needs to be readily accessible for calculations and analysis?
- How fast do the calculations have to be done?
- What systems exist for coding project information? If new coding systems will be developed, what constraints (e.g., number of digits) may be imposed by existing practices?
- How much flexibility is required in the system's operations? Which operations need to be flexible and to what extent?
- What cost limitations will apply to the design and implementation of a computer system? What schedule limitations apply?
- Of all the specified features, what is the priority of each one? For example, is price more important than storage capability? Do we prefer buying a software package "off the shelf" or designing our own? Can we separate our "wants" from our "needs" in terms of system requirements?

Specifying Long-range System Requirements: Specifying requirements requires a reevaluation of the issues outlined in our discussion of short-range requirements, but with a view toward the future. Any system we develop must be able to meet future as well as current requirements or be adaptable to newer systems as they become available. For example, if the workload of the project engineering department is expected to double in the next two years, we must be sure that our system is expandable to that capacity. Because the company may change or add to its computer systems, any systems we introduce will have to be compatible with the new system. New codes of accounts, procedures, and interfaces may also be implemented.

One key aspect of system design is the identification and evaluation of "wants" versus "needs." As in any design function, there is a tendency to include capabilities that are desirable but not really necessary. The problems of distinguishing between wants and needs is particularly difficult in system design, since so many of the benefits have to do with "managing better" and are therefore difficult to quantify. One way to keep the system's specification under control is to specify the minimum system that will do the minimum job and then put each added feature or capability through a cost/benefit analysis just as one would do for a proposed design change.

Step 2: Developing System Performance Specification

Step 1 was aimed at determining what we want the system to do for us. Step 2 involves describing the specific characteristics of the system which will have the capabilities we need, both now and in the future. Performance specification consists simply of describing the four basic system elements: input, processing, storage, and output.

Specifying System Input: To describe system input, we must specify the type of input, the amount of data to be inputted, the sophistication of the person making the input, the location of the input function(s), the flexibility required, and the need for interaction with the main system during input. The input to the system can take the form of interactive terminals, electronic data transfer systems such as disks or tapes from other computers, or through electronic communication.

If input is to be in the form of interactive terminals, which fits most project environments, we have a number of options:

1. "Smart" terminals which are actually small personal computers that provide local computing capability. Such terminals can process input data so that it goes into the main system in a certain format and the user can perform certain functions at the terminal without having to access the rest of the system. When the terminal performs some processing and then inputs the result to the main system, it is said to be *uploading* information into the system. When the main system performs some processing and then sends the resulting information to a terminal, it is said to be *downloading*.
2. "Dumb" terminals, which are strictly input/output devices for the main system.
3. Local area networks (LANs), in which a number of computers and software packages are linked together and are able to interact with each other.
4. Menu-driven input and/or command-driven input.
5. Remote terminals with tie-in to the main system via telephone, telegraph, or other forms of telecommunication.

An example of input by data transfer is the use of timesheet data which is already input to the accounting system. We might arrange for the workhours in each labor category to be input to our cost control system.

Specifying System Processing: In this step we must specify what the system must do with the input data: that is, what calculations are to be performed, how much data needs to be handled, how many calculations must be made, how quickly the calculations must take place, what data files will have to be read, what comparisons will be necessary, and what results are desired. It must also be recognized that an interactive system allows the user to follow and even to affect the data processing function, and we therefore need to identify those calculations which are always the same and require no interaction and those for which interaction is desirable. For example, the preparation of weekly expenditure reports would, in general, not be an interactive process, whereas the routines used for decision-making would be. Therefore we need to specify the operations that are to be interactive and the extent of interaction desired.

The system processing functions that are most often used in project management are arithmetic calculation, data retrieval, comparison, and extrapolation.

Calculation functions are performed the same way we would prepare a progress report, an estimate, or any other simple task. Much of the work involved in progress report preparation is the calculation of current figures and the comparison of those figures with the budget plan. To make the calculations by computer we retrieve the file of budget data and perform a comparison of actual data versus the budget plan. The extrapolation of current trends is a useful way to highlight potential problems; it is another function that is easily computerized.

These functions are generally provided by the purchased software. One feature to look for in software is the flexibility to define the functions that are to be performed. For example, there are many scheduling packages that do network analysis, but they all vary in terms of the user's flexibility in defining the coding system, selecting arrow or precedence notation, defining subprojects, loading resource and cost data onto the activities, and so on.

Another important feature to look for in software is the ability of one software package to interact with other software. For example, in cost engineering we are interested in network and resource analysis, cost estimating and forecasting, and management reporting. We may also wish to use other software categories such as cost estimating, risk analysis, spreadsheets, word processing, management graphics, and databases. Although most systems have software to do each of these functions, we may find that the ability to tie them all together will vary from system to system. If these functions can be integrated, the result will be a much more efficient total package.

Specifying System Storage: We must specify the amount of data to be stored, the format for storage, and the requirements for recall. For example, we might wish to construct a database of norms for estimating derived from past projects. This database would be carefully designed and probably integrated with an estimating routine. We might also wish to store, retrieve, cost, schedule, and design data from past projects for future analysis.

Specifying System Output: The success of system implementation depends, to a great extent, on the effectiveness of the output. If a system produces clear, concise, useful, and easy-to-read charts and reports, it is bound to be a success. Conversely many good systems fail to win the acceptance they deserve because the users simply cannot relate well to the output produced. This principle is particularly important in the project environment where we are trying to present "the right information, in the right format, to the right person, at the right time."

There are three types of output which are relevant to all projects: printed output, visual displays, and graphics.

Printed output

The reports generated by printed output will probably be circulated throughout the company to advise managers of status and problem areas. These reports should permit the practice of management by exception and provide only that information which is necessary to the recipient of a particular report. In order to target the appropriate audience, the system can include a feature called the *flexible report writer*. This feature enables the design of many reports, each of which has a separate format and which displays specific data. Systems without this feature have one or more report formats that cannot be changed.

Another useful feature in report preparation is the ability to sort, select, and order data. This option makes it possible to select the specific information we want for each report, sort it into appropriate categories, and print it in a given order. For example, a manager's report listing all projects might be designed such that the project with the largest cost overrun is printed first, and subsequent projects are ordered by decreasing overrun amounts. From our array of project data we might select cost data for an accounting report and order it by cost code; the same data might be arranged by contract number for a report to the contracts engineer.

Visual displays

The image that an interactive system makes on the screen is, of course, important only to the user. However visual displays are an important form of output because the user's effectiveness depends to a great extent on the system's ability to give quick outputs of useful information. One useful feature is the ability to present several displays on one screen. Another feature to look for is the display of messages that tell the user what is going on when the computer is calculating and also what is going wrong. Finally we would like to have a visual display that makes it easy to see the results of a change to the project parameters. For example, we might be varying logic and durations to test the effect on the end date. We would therefore like a display format which makes it easy for us to see the end date after each run.

Graphics

Nowhere is the principle that "a picture is worth a thousand words" better illustrated than in management graphics. With well-designed graphics, the meaning of an otherwise confusing and uninteresting set of figures can be driven home clearly, quickly, and forcefully. Because a great deal of the effectiveness of our cost control effort will hang on our ability to focus the attention of busy managers to those areas in which it is required, management graphics deserve consideration in any project environment. Most computer systems developed for project management have such a capability.

To specify the graphics capability required, we need to identify the type of graphics, the quality required, the source of data to be used, and the quantity required. For project and cost engineering applications, the most common graphics types are:

- Barcharts
- Networks
- Labor histograms
- Organization charts
- Pie charts
- Tracking and trending curves

The quality determines the type of drawing equipment required. For example, if top quality color graphics are desired, then a four- or eight-pen plotter with associated software is needed. Plotted management graphics are the best possible way to communicate plan and schedule information, and they are strongly recommended. These may often be created by exporting project data to a spreadsheet or graphics program.

Step 3: Preparing the Plan and Cost Estimate for Implementation

At this point we have an idea of what we are trying to accomplish with computerization and what kind of system we need. Before going further it is appropriate to exercise some cost engineering principles ourselves and establish a plan and estimate for what is to come.

The major cost items that can be expected are as follows:

- Purchase or lease of hardware (i.e., computer and peripheral equipment)
- Purchase or lease of software
- Purchase of accessories
- Maintenance contract for hardware and/or software (if applicable)
- Modifications to office facilities to accommodate computer (if applicable)
- Time required for training and implementation
- Consulting services for applications, training and support (if applicable)
- Supplies (computer paper, disks, etc.)

The computer acquisition and implementation should, in fact, be treated as a project just like any other, with a budget and a schedule. It should be recognized that the estimate prepared at this stage will be preliminary in that we have not yet gone into the marketplace and obtained quotations. However it is important to prepare a preliminary budget at this stage in order to accomplish the important next step—obtaining management approval to proceed.

Step 4: Obtaining Management Approval to Proceed

The purpose of this step is to advise management of what we are doing and obtain their general approval of the overall plan and their specific approval to proceed with system evaluation and recommendation. This step is suggested for three important reasons:

1.   We wish to have company management "on our side" in this endeavor, because the organizational implications of a new computer system tend to be far greater than the capital cost might indicate. We want to be sure that management understands and supports the objectives of better project management toward which the computer is merely a means. We can point out that the benefits of formal cost engineering and project management methods and systems include:

- Reduced project risks
- Increased consistency between projects
- Improved decision-making
- Greater accountability
- More flexibility
- Increased productivity
- Cost effectiveness
- Greater responsiveness and more

2.   We wish to impress upon company management that the computer system is intended to be a cost-saving device that is to be evaluated and managed like any other small project. Far too often engineers and managers decide first that a computer is needed, then decide what kind of computer to buy, but never submit these subjective decisions to the same sort of scrutiny that any other investment decision would endure. We must never lose sight of the fact that there is only one good reason for computerization—that it will be profitable. We want our management to know that we are operating that way.

3.   By providing preliminary cost/benefit figures, we are in a position to review our plans with management and perhaps identify some changes. For example, we might be pleasantly surprised to find that management is enthusiastic about the project, which encourages us to specify and implement a more ambitious system than we would have done on our own. And, because cost engineering affects so many parts of the organization, there may be implications and interfaces that only management are aware of, and that insight is an important part of a successful system.

Once we have obtained approval to proceed further, we are ready to journey into the world of modern computing. But before we are swept up in electronic excitement, we need to prepare ourselves for this journey. As always, the best way to be prepared is to know exactly what we want.

Step 5: Preparing the Sample Project and Reports

Although we went to a great deal of trouble in Steps 1 and 2 to define our needs, we had better expect that there will be many systems that will meet or exceed them. We had also better expect that we are likely to see a bewildering array of computers, salespeople, and impressive demonstrations. At the end of our search we are likely to say, "But they all seem so good. How can I possibly choose one?"

The truth is that all the systems are probably very good. Each has features which its proponents claim make it immeasurably better than the others. The only difference between them will be that some will suit our needs better than others, and one will suit us best of all. But how will we know when the right one comes along?

The answer is to prepare a sample project (or projects) that represents the typical situation which our system will have to handle. We should also make up some input data in exactly the form in which we would normally receive it and design some management reports that show exactly what we would like the system to produce. Now, instead of approaching our friendly computer salesperson and saying, "Show me what your system can do," we can say, "This is what I need; show me that your system can do it!"

We have therefore defined system requirements, system performance specification, and a sample project for use in testing and evaluation. We know what we want and have expressed it in specific terms. At the same time we are free to change any aspect of our system design as we become aware of new system features and capabilities. We have also established a basis for fair testing and comparison of the systems selected for in-depth evaluation.

Step 6: Contacting Suppliers and Developing a Short-List

The purpose of this step is to identify three or four systems which are worthy of in-depth evaluation. When conducting this "survey" of what is available, bear in mind these few hints:

1. *Consider software before hardware.* The effectiveness of a system will be determined first and foremost by the software. If the programs meet our needs, then it is likely to be relatively simple to arrange the hardware as necessary.

2. *Take the time to read the user's manual.* Short of running the software there is no better way to get a feel for what the software can do than to read the user's manual. The clarity and completeness of the manual is itself a good indication of the quality of the software package.

Unfortunately, as a cost-saving device there is now a tendency for software providers not to provide a user's manual. Instead they rely on a built in tutorial and/or a help menu. In such cases, a demonstration copy of the software and the tutorial should be requested from the supplier. If this cannot be obtained, it is a good indication that the software may not be adequate.

3.   *Avoid writing new software.* Many people jump too quickly to the conclusion that their needs can only be met by writing their own software. This situation may be true in some cases, but in most cost engineering/project management applications there are few functions which cannot be satisfied by some existing software package. Although the existing package may represent a compromise, it must be remembered that the time, cost, and risk are greatly reduced. Software development projects are notorious for overrunning budget and schedule by several orders of magnitude.

4.   *It pays to be skeptical.* For example, one should pay no attention to claims that a certain capability we require will be available "on the next release of the software." We are interested (at least initially) only in that which is available and demonstrable now.

5.   *Insist on talking with someone who is knowledgeable in the specific area of interest.* With the proliferation of computer stores and companies, combined with the proliferation of hardware and software products, we could hardly expect an equal proliferation of experts. It is, of course, impossible even for a real expert to be familiar with every application of every system and, as a result, good advice is hard to find. So, when discussions progress to the point where it is possible to identify the items in the hardware or software product line which are of interest, it is worth the extra trouble to find and talk to the person who is most knowledgeable about those specific items.

6.   *Examine closely any claims about hardware and software interfaces.* Experience has shown that the problems of getting machines to communicate with each other and the problems of linking software are often far greater than anticipated.

At the end of this step we should have identified three or four systems for in-depth evaluation. More candidate systems will only add to the time and cost of the evaluation process and will probably not increase the likelihood of the correct system being chosen. During this step we may also have made some revisions to our system design to reflect what we have learned. Now we are ready for the "try before you buy" part of the project.

Step 7: Conducting the "Compute-Off" and Evaluating the Results

The purpose of this step might be compared to selecting a new car by driving each car selected for possible purchase around the same course and evaluating the results according to a set rating sheet. In this case our "course" is the sample project, data, and reports.

Acquiring the Software: We begin by acquiring a full copy or a demonstration copy of each of the candidate short-list programs. The low cost of personal computer software usually makes purchasing a viable route. We then proceed to learn each program and to use it to test its ability to plan and analyze the sample project. We test its capabilities against our software specifications.

Preparing the Rating Sheet: The categories on the rating sheet and the relative weight given to each one are determined by our definition of the system's design requirements. However we can identify some system characteristics that will inevitably be important for a project application.

1. *User-friendly interaction.* The system should be easy to use, with a low "frustration factor." User-friendly features include English-type commands (as opposed to letters or numbers), tutorials (in which the system actually guides the user along), helpful error messages (e.g., "system will not accept decimal input" is more helpful than "syntax error") and help commands (when in doubt, the user simply types "help" for a complete listing and explanation of the commands available at that point).

2. *Flexible reporting format*, or a fixed format that is well-suited to the particular application.

3. *Good interface characteristics between the hardware and any existing hardware in the company.* For example, it is a plus if our new microcomputer is capable of communicating with the existing mainframe computer even if we have no intention of doing so at present.

4. *Good interface characteristics between software packages.* For example, can we pass data and information from one software package to another?

5. *Good user support* from the hardware and software suppliers.

6. *Software documentation* providing a clear, easy-to-use user's manual.

7. The provision of *training programs* for users.

8. The *ability to upgrade* hardware and/or software to increase capacity or to use newer products.

We should bear in mind that we are rating the software, the hardware, and the supplier both in terms of price and performance.

Evaluating the Results: To evaluate the results we like to use a small committee of individuals with diverse skills, such as computing, project control, estimating, accounting, and management reporting. Each individual rates each system, and a composite score is agreed upon by the committee. Evaluations should be made, at first, without regard to price. Once the performance ratings are established, then price can be considered and cost/benefit analyses performed.

Step 8: Obtaining Management Approval for Implementation

The purpose of this step is to present a specific proposal for implementation to company management and to obtain approval for full implementation. As before, it is important to treat the subject of computerization objectively and show in specific terms how profitability will be enhanced, due to time and cost savings on projects and in the project engineering and management functions. It is important that the first step in computerization be one which provides visible and significant benefits so that the concept is well accepted.

Management support is even more important at this stage than it was when obtaining approval to proceed. During implementation people are likely to react with cautious enthusiasm, tolerance, grudging acceptance, reluctance, or even outright hostility. We will be much better equipped to handle this if we and everyone else knows that company management supports the program.

The proposal to management will probably be an update of the preliminary proposal that provides more details on the time, cost, and plan for short range implementation, a summary of the possible longer-range developments, and a more detailed economic justification. The organizational aspects of computerization will be uppermost in the minds of management. That issue should also be squarely addressed by describing how the roles of existing functions may be changed, how lines of communication will be altered, and how efficiency in various departments will be improved.

Step 9: System Implementation

Setting Up the First Project: After installation of our computer system, the next question is where to start. Many companies find it useful to take a project and implement it on the new system while continuing to control it using the old methods. This provides a check on the results and avoids making a "guinea pig" out of the first project. Often there are lessons learned from the early projects that indicate improvements that can be made before the system is fully implemented.

Keeping Everyone Informed: A key factor in the acceptance of the new system is that everyone must be kept informed, and we must display a willingness to share experiences with the new system and to help others obtain similar benefits. It is especially important to be sure that those individuals and organizational functions that interface with our new system are made to feel that they were consulted early enough to assure that their needs and constraints were considered.

Some of the methods for keeping people informed include progress memos, coordination meetings, and in-house seminars and training sessions.

Preparing Procedures: It is important to assure that the new system is used in a manner that is consistent with all the project's engineering and management functions. The best way to ensure consistency is to prepare procedures that specify how data are inputted, processed, and outputted. The procedures should show how each organizational function interfaces with the system and should specify who is in charge of the system. Each function that will use the system should have specific procedures to follow.

Many systems have different levels of security that are protected by passwords and/or other procedures; therefore the limits of each user must be defined. For example, payroll or personnel data can be included in the system, but access to those data must be restricted. Similarly we might construct an estimating database that is available to any cost or project engineer for estimating work but can be updated only by those responsible for it. In that case access to "read" from the database is unlimited, but access to "write" to the database would be restricted.

The procedure should also specify who the "computer czar" is to be, that is, the person who answers questions, resolves problems, sets priorities, listens to complaints, deals with the vendors, and generally coordinates system use. The steady expansion of the use of personal computers in business has had the somewhat undesirable side effect of diluting the supervision and coordination of computer usage within a company. In past years companies purchased a mainframe computer to service the entire organization, and users had to obtain the services they needed from the data processing (DP) department. Because systems were expensive, maximum utilization was the goal, and users were often faced with a long wait for the needed capability to be provided by an already overworked system. The microcomputer has enabled users to implement their own solutions in their own way without recourse to the DP department. Although this change has broadened the effectiveness of computers, it also has created many situations in which individual effectiveness has improved, but organizational efficiency has not.

Therefore today's responsible user must assume some of the responsibility previously borne by the data processing department to make sure that others are aware of what they are doing and that their system is compatible with others in the organization.

Training: Formal training courses are a significant aid in the acceptance and effective utilization of a computer system. The training course can be based on the user's manual (which describes how to operate the system) and the procedures (which describe how the system fits into company operations).

11.5 A REVIEW OF PROJECT CONTROL SOFTWARE

Because most project control needs can be satisfied by a variety of existing systems, it is worthwhile reviewing some characteristics of current systems. Software can be divided into a number of categories as discussed below.

11.5.1 Planning and Scheduling Software

A great many software packages are currently available to perform planning and scheduling. Virtually all perform forward and backward pass calculations of critical path and float. Some features that vary among software packages are:

- The size of the project handled (i.e., number of activities)
- The ability to use precedence or arrow notation
- The flexibility of the coding system used to identify activities
- The ability to create and use subprojects

- The ability to calculate free float
- The ability to perform resource allocation, resource aggregation, and resource leveling
- The number, type, and flexibility of resources permitted
- The flexibility of displays and graphic output
- The speed of analysis
- The ability to handle multiple projects simultaneously
- The ease and flexibility of updating (e.g., out-of-sequence updating that allows updates even if activities have not been processed in strict logical sequence)
- The number of users permitted
- The amount and flexibility of the data that can be attached to each activity

The selection of a package will depend, of course, on the needs and priorities of the user.

11.5.2   Cost Estimating Software

Cost estimating packages are variable in their applicability to different types of projects. The reason could be that the approach taken to cost estimating is less standardized than is planning and scheduling, making it more difficult to create a standard package that would satisfy all users.

A number of available software packages are designed for specific types of estimating, such as architectural work or industry projects. If an available program suits the particular application, all is well and good. However there may not be a program that exactly fits the user's application. Fortunately many available software packages are suitable for general estimating applications. If we consider that cost estimating usually consists of much multiplication of quantities by unit costs and addition of the results, it is evident that it is a simple matter to computerize it. The methods and data required for estimating can be developed with available software. Two useful types of software for cost estimating are spreadsheet and database programs.

Spreadsheet Software: Spreadsheets enable the user to define a matrix of data as well as the calculations required to define additional data. One can, for example, take a column of quantities and multiply it by another column of unit costs to create a detailed estimate. Spreadsheets for detailed calculations are then summarized and subjected to further manipulation in a higher level spreadsheet.

Database Software: Cost estimating applications are designed to collect, analyze, normalize (i.e., adjust to a consistent basis or "norm"), and present the basic data needed for estimating. Because the best predictor of future costs is usually our experience, and small projects continuously generate a lot of useful, actual cost data, a database is a handy way to organize it all.

Databases and spreadsheets can be integrated with planning software in order to automate the estimating process described in Chapter 5, in which the labor and construction equipment resources are estimated from a standard file of rates. This process makes it possible to examine the cost impact of different planning scenarios as well as to make cost forecasts reflecting current progress.

A cost estimating database should have an independent variable (e.g., design parameters such as cubic yards of concrete), and a corresponding value for the dependent variable (cost). It should be organized according to cost code so that data can be easily selected, sorted, and summarized. Other database applications are discussed below.

Out of the many database packages available, some are intended for use in conjunction with a network package. The database, as can be seen from the cost engineering applications described above, enables us to collect, organize, analyze, sort, select, compare, and present any kind of data in any way we want. Project management applications of databases, in addition to estimating, include:

- Data on the durations of typical design and construction activities
- Material control by storing and accessing data on purchase orders, purchase requisitions, delivery dates, quantities, etc.
- Document control (i.e., keeping track of drawings, specifications, and critical communications)
- Vendor data
- Contractor performance data
- Maintenance data (i.e., required inspection and maintenance intervals and procedures, and actual maintenance records for each item)
- Weight data (for those projects in which weight is critical)
- Personnel data
- Wage rate data

Some database programs are like blank slates and can be designed in any way that the user requires, while others are designed for a specific application.

A very useful feature of some systems is the integration of a database with the network analysis program. This feature enables us to identify a resource requirement for an activity and link that resource to a detailed set of relevant data in the database. For example, we might identify heat exchanger E103 as a resource requirement for the activity "erect heat exchanger." By setting up a link to our database we can have the program select some or all of the data pertaining to E103 and advise us of such facts as the vendor, the purchase order number, the price, the promised delivery date, the current status, and the current delivery date. We can even use this capability to update our schedule automatically to reflect current deliveries or to update our cost forecast automatically to reflect current progress.

11.5.3 Economic Analysis Software

Many available software packages perform economic analysis in order to evaluate project profitability. These programs can be useful in establishing the return on investment or in monitoring the projected return as changes occur in the original basis. It is worth noting that the profitability analysis is seldom updated after the project is approved, although it should be, as it is a relatively simple task for the project engineer to use an economic analysis to make decisions based not on minimum cost or schedule, but on maximum profitability.

11.5.4 Management Reporting Software

Management reporting software consists of word processing, special report generating packages, and management graphics. This software may be available within the existing computer system of the company. Features to look for in these packages include the ease of use, compatibility with existing systems (such as word processing), and the ability to interface with the main project management programs. It is highly desirable that the report writing and management graphics packages be either part of or fully compatible with the planning, scheduling, cost estimating, and project control package(s) to avoid duplication of input and to assure timely reporting.

11.5.5 Risk Analysis Software

For projects involving significant financial risks, a risk analysis program may be helpful in addressing the probability of various costs and schedules being achieved (see Chapter 9). For example, bidding strategy may necessitate an evaluation of bid price versus the probability of getting the job. Or a turnaround project that incurs major costs for every day the unit is shut down may benefit from a probabilistic analysis of the schedule to determine the probability that the unit will be running on a given date and enable managers to plan accordingly.

Probabilistic analysis is appropriate for many more project management situations than is generally acknowledged, probably because many engineers and managers are somewhat suspicious of using statistics as a means of decision making. However, a number of available software packages do a credible job of analyzing the probability of cost overrun or schedule delay. In general these programs evaluate costs separately from schedules and use Monte Carlo simulation to generate the results. In Monte Carlo simulation a random-number generator is used to select a value for each activity duration or cost variable until a sample cost or schedule outcome is calculated. By repeating this process hundreds or even thousands of times, a probability function describing all possible outcomes can be developed. The major drawback to this type of analysis is that it requires that the variables be independent—in reality, they seldom are. However most programs provide a method to adjust for this assumption of independence.

If risk and contingency analysis is important, an appropriate software package can make the inevitable judgments more rational, credible, and acceptable. And, an appropriate package can point out the magnitude of risks that were previously thought to be inconsequential.

11.6 COMPUTER-AIDED DESIGN APPLICATIONS FOR PROJECT MANAGEMENT

The trend toward greater computing power at lesser cost has been very evident in the area of computer-aided design (CAD). As a result both owner and contractor companies in many fields are finding that CAD offers great improvements in the efficiency with which design tasks are conducted. CAD systems also offer the potential for dramatic improvements in cost engineering/project management functions as well.

11.6.1 Basic CAD Components and Capabilities

A CAD system is designed to perform many functions associated with the design process:

- Drafting, including layouts, revisions, and final drawing
- Storage, recall, and display of relevant design information
- Material takeoffs for procurement
- Modeling in three dimensions to remove impediments or space conflicts
- Performance of design calculations
- Application of design standards

To accomplish these tasks a CAD system typically consists of the following components:

1. A workstation consisting of one or more screen displays, and an electronic table, a keyboard, and a device for locating points on the screen or on the table (e.g., "mouse"-type cursors, light pens, "joysticks," thumbwheels, etc.)
2. A central processing unit providing the hardware and software to perform the necessary operations
3. Storage facilities, such as tape or disk
4. Auxiliary input devices (as required)
5. Telecommunications facilities (as required)
6. Printers and plotters to prepare the drawings

In most cases a single processing unit and suite of output devices can support a number of workstations. As a result of the efficiencies introduced by CAD, both owner and contractor companies have experienced improvements in design efficiency by a factor of 4 or more. CAD systems are of interest, not only because their use is widespread but also because they have some features that are of great interest to project engineering and management functions. Foremost among these features is the database capability, enabling us to attach management data to design graphics. Current technology also embraces computer-aided manufacturing (CAM), in which the design of a part to be manufactured can be translated directly to a tape for a numerically controlled production machine.

11.6.2 CAD Applications to Cost Control

One of the most common and long-standing challenges in cost engineering involves the exertion of cost control during the design phase. Typically this effort takes the form of procedures to control design workhours and costs as well as to estimate and track design changes. However the most important aspect is usually not addressed—control of cost implications of design decisions. This is vital because we have the most control over the design aspect of the project, and design is the aspect that has the greatest potential impact on cost.

Unfortunately most companies are not very successful at integrating design and cost engineering. This weakness is partly due to the misplaced belief that engineering work should be done with minimum time and cost. This criterion, of course, discourages any discretionary work such as efforts to reduce cost through design changes.

The CAD system enables us to construct a database for estimating the cost of each component included in the design. Thus the cost of each alternative design configuration can be easily calculated and provided to the design engineer. Changes can be identified and provided to the design engineer, and can also be identified and estimated more efficiently. Life cycle costs can be integrated into the design process.

11.6.3 CAD Applications to Weight Control

Many projects, such as offshore platforms, require close monitoring and control of weights. Control can be handled in a manner similar to that used for cost control by constructing a database of weight information attached to the graphical representation of each design component. The resulting data and analysis of weights provides vital information for platform structural analysis and planning of construction operations.

11.6.4 CAD Applications to Document Control

All projects require storage and organization of documents, not only for efficiency but also for certification and operation of the facility. Since many documents relate specifically to a design component, the CAD system can be used to log in all documentation relating to a specific item of equipment or type of bulk material. Typical documents stored in this way include:

- Quality control documentation, such as testing and inspection results
- Operating information

- Vendor information
- Maintenance information

11.6.5   CAD Applications to Maintenance

A typical problem of engineering and construction projects, large or small, is the problem of "handing over" the completed facility to those who will operate it. The operations personnel are interested in data that have (presumably) accumulated during the engineering phase but was of little interest to those who were designing and building the facility. Consequently the operations staff often finds it difficult to obtain and organize the information needed. One example of this problem is maintenance.

In order to assure that an effective program for inspection and maintenance is set up, we can use the CAD database to accumulate information relating to inspection and maintenance of each item shown on a drawing. The database could show, for example:

1. The frequency of inspection
2. The intervals between planned maintenance
3. The procedures to be followed for planned and unplanned maintenance

Once the facilities are in operation and the maintenance program is in effect, additional information can be added to the database to facilitate planning and monitoring of the inspection and maintenance functions. Specific items include:

- The date of last inspection
- The result of the last inspection
- The date for the next inspection
- The date of the last planned maintenance activity
- The cumulative maintenance performed
- The date of the next scheduled maintenance
- The specific maintenance item(s) scheduled

11.6.6   CAD Applications to Construction Planning

Most CAD systems have the capability to work in three dimensions. This capability makes it possible to model the construction work to be done and see potential interference, misfits, or impractical construction requirements. For example, one use of a plastic model is the ability to see what is going to be built in three dimensions. The three-dimensional model on a CAD system not only lets us see the project in three dimensions but also enables us to "move things around" with ease. This feature can be used to test the feasibility of different arrangements and to actually simulate construction operations such as major lifts. For small projects in dense operating plants, the three-dimensional modeling feature can be a big help in avoiding expensive situations in which the design looks good on paper but proves to be impossible to construct in the field.

Finally, by simply speeding the design process and the production of approved drawings for construction, the CAD system can do a lot to make construction go more smoothly.

RECOMMENDED READING

Drigani, F. (1989). *Computerized Project Control*, New York: Marcel Dekker, Inc.
Samid, G. (1990). *Computer-Organized Cost Engineering*, New York: Marcel Dekker, Inc.
Westney, R. E. (1992). *Computerized Management of Multiple Small Projects*, New York: Marcel Dekker, Inc.

12

Cost Engineering Ethics

12.1 BLACK AND WHITE, OR GRAY?

In the May 1998 newsletter of the North Carolina State Board of Registration for Professional Engineers and Land Surveyors, the board chairman, R. Larry Greene, said:

> When I was a young man, acting in an ethical manner seemed to be a straightforward proposition. An act was either right or wrong, black or white, morally correct or [not] ... As an older, perhaps wiser, man I still try to guide my life by those principles, but I have learned over time that recognizing right, wrong, black and white is not always a simple proposition... Every situation we find ourselves in is different from any other we have ever experienced. As each new scenario is added to a job situation, the ethical mix changes and the boundary between ethical and unethical conduct, perhaps already gray in nature, shifts.

On the other side of the world, in the June 1997 issue of *The Building Economist*, the journal of the Australian Institute of Quantity Surveyors, Tony Fendt, managing director of Project and Retail Pty. Ltd. said:

> We work in an uncertain world; an increasingly global economy where the competition for employment increases by the day, and the pool of opportunity decreases by the day whether due to population increase, advancement in technology, downsizing, increased per capita production, etc.

This chapter is excerpted in part from the book *What Every Engineer Should Know About Ethics*, by Kenneth K. Humphreys, Marcel Dekker, Inc, New York (1999) and selected articles from a series entitled "The Ethics Corner" authored by Kenneth K. Humphreys that appeared in *the Professional Engineer: The Magazine of North Carolina Engineering* beginning in 1996. The series was subsequently reprinted in *Cost Engineering* magazine.

The question is are we justified in stretching the truth, or even conceal-
ing the truth, in order to gain or retain favour with a client?
In order to answer this question, we need to examine our moral roots.
Ethics is primarily concerned with what we ought to do when what is
right and what is advantageous and profitable conflict with one another.

Mr. Fendt goes on to quote Cicero's letter "On Duties" written to Cicero's
errant son, Marcus:

Let us regard this as settled: what is morally wrong can never be advantageous, even when it enables you to make some gain that you believe to
be to your advantage. The mere act that believing that some wrongful
course of action constitutes an advantage is pernicious.

Ferenz (1999) has defined ethics as:

...what we ought to do and how we ought to behave from a moral
viewpoint, as opposed to an economic, religious, political or prudential
viewpoint.

These various comments illustrate the clear relationship of ethics to moral-
ity and also suggest that the definition of ethics is strongly dependent upon
current cultural and social viewpoints. What is obviously ethical and moral in
one context may be quite the opposite in another. In engineering, the distinction
between black and white is often even less clear. Engineers operate in a multinational, multicultural business environment in which what is considered moral and
ethical often varies from one location to another. In some cultures it is legal and
totally acceptable morally for substantial gifts to be exchanged between those
desiring to do business in that area and those seeking to have work done. In
Western cultures, this is considered to be bribery and is illegal.

Ethics clearly is not black or white—it is many shades of gray depending
upon the given situation. Nevertheless, engineers do have guidance in determining what ethical standards they should apply to their life and work—the codes of
ethics of their professional societies. These codes define for the engineer what is
acceptable and what is not. They define what engineering ethics is, and what it is
not. Every engineer and engineering student should become thoroughly familiar
with the code of ethics of his or her disciplinary engineering society. These codes
form the foundation for ethical practice and, not coincidentally, generally are
written in whole or in part into the laws and regulations governing the practice
of engineering in the United States.

12.2   THE INTERNATIONAL COST ENGINEERING COUNCIL CANONS OF ETHICS

In 2000, the International Cost Engineering Council (ICEC) adopted a set of
"Canons of Ethics for Cost Engineers, Quantity Surveyors, and Project Managers."

Recognizing the different ethical criteria that apply around the world, ICEC did not believe that it could adopt a detailed Code of Ethics such as that of the National Society of Professional Engineers (NSPE) or any of the disciplinary engineering societies in the United States. ICEC instead adopted a very simple set of ethical canons based upon the "Preamble" and "Fundamental Canons" of the "Code of Ethics for Engineers" of the National Society of Professional Engineers (NSPE). The ICEC Canons are brief but provide, in a limited number of words, a broad set of ethical guidelines. The full text of the ICEC Canons follows:

ICEC CANONS OF ETHICS FOR COST ENGINEERS, QUANTITY SURVEYORS, AND PROJECT MANAGERS (ADOPTED BY THE ICEC EXECUTIVE, JULY, 21 2000)

Cost engineering, quantity surveying, and project management, are important and learned professions. Members of these professions are expected to exhibit the highest standards of honesty and integrity. The services provided by cost engineers, quantity surveyors, and project managers require honesty, impartiality, fairness and equity, and must be dedicated to the protection of the public health, safety, and welfare. All members of these professions must perform under a standard of professional behavior that requires adherence to the highest principles of ethical conduct.

Cost engineers, quantity surveyors, and project managers, in the fulfillment of their professional duties, shall:

1. Hold paramount the safety, health and welfare of the public.
2. Perform services only in areas of their competence.
3. Issue public statements only in an objective and truthful manner.
4. Act for each employer or client as faithful agents or trustees.
5. Avoid deceptive acts.
6. Conduct themselves honorably, responsibly, ethically, and lawfully so as to enhance the honor, reputation, and usefulness of their professions.

12.3 COMPUTER-AIDED NIGHTMARES

Chapter 11 discussed at some length computer applications for project control and cost engineering and some of the software not readily available for cost and schedule work. Chapter 11 also stressed the need to fully understand what any given piece of software can do and what its limitations are.

All too frequently today, software is misused or is not suitable for the job to which it is applied. Today engineers rarely write their own software and, as a consequence, often do not fully comprehend what the software is capable of doing and what its limitations are. The result is often disastrous and, unfortunately, has led to project failures, structural failures, injuries and loss of life. To use a piece of software without fully understanding its capabilities is foolish, can be dangerous, and is unethical.

Years ago, off-the-shelf software was something you only dreamed about. There was no cadre of readily available programmers, and there was no convenient computer supply store where you could buy a software package to do virtually anything. There was only yourself and your ability to write your own programs. For an engineer that usually meant mastering FORTRAN (FORmula TRANslation), a computer language created for engineers to convert our mathematics and formulas into something the computer could understand.

You painstakingly analyzed the problem at hand, decided the method of solution, sat down and wrote out the mathematics and formulas, and then converted that to the FORTRAN language. Next you punched it all into a stack of cards and read them into the computer which hopefully did what you wanted, but more often than not didn't. You usually had a long sequence of try, debug, try, debug, and try again before you finally had a workable program which did what you wanted. It was indeed a tedious proposition!

However, when you did finish writing and debugging the program, you knew what it contained. You knew what it assumed, what engineering principles applied, and when it was valid to use the program. You also knew when the program was not valid. It was an extension of yourself—it contained your engineering experience, judgment, and expertise.

Today, we are far beyond that. Engineers rarely write their own programs anymore. We have programming staffs to do that or, more likely, we buy a software package from one of the many very qualified suppliers in the business. A vast improvement isn't it? **Not necessarily so!**

One of the problems with software obtained from others is that, no matter what the documentation or salesperson says, you can't be sure that it will always apply to your situation. You don't necessarily know the inherent assumptions which went into the design of the software nor do you know with certainty what its limitations are. These computer programs and the computers themselves are wonderful tools, **but don't use them blindly**. Question what the software produces. Don't take the output on faith. It might not be correct. The software might assume "average" conditions, but your particular project might deviate from average (and probably does).

Your name goes on the project, not the name of the anonymous person who wrote the program. If the design is flawed, you are responsible—not the programmer. You have the professional and ethical responsibility to check the results. Examine them. Are they reasonable? Do they fit the project conditions? Do they make sense? Do they coincide with what your experience and judgment suggests? They should, and if they don't, find out why.

An impressive engineering report prepared by a major U.S. engineering and construction (E & C) firm not long ago just didn't ring true to a reviewer. His experience waved a big red flag so he asked to meet with the engineers who had prepared the report. He learned that virtually everything in the report had been generated using a very widely known and respected piece of software, a software package which is generally accepted as being very reliable. *This time it wasn't.*

Why? The package assumed a certain set of design conditions for a project unless the input specified that different conditions applied. Unfortunately the documentation did not clearly explain the default assumptions built into the software. In this project, many things were different, but the input had not been modified accordingly. The result was that the report was totally flawed—the program simply was not applicable to the project at hand. The E & C firm should have known this but didn't. They were comfortable with that software and had grown complacent in using it. The engineers assumed it was valid. They didn't check the results for reasonableness, and they got burned.

That E & C contractor lost the job, one which was worth over $500 million. In this case, blind faith in a piece of software cost the firm a lot of money. In another set of circumstances, this type of error could have cost lives.

12.4   A COMPUTER-ASSISTED CATASTROPHE

Prior to the advent of the digital computer, engineers used slide rules for design calculations. The slide rule was the symbol of the profession and engineering students at any university could usually be visually identified. On their belts they wore a leather holster containing a slide rule. The most common slide rules were about ten inches in length, and no engineer would ever be without one. They were vital tools of the profession.

A popular slide rule was a Keuffel & Esser "Log Log Duplex Decitrig." It was a beautiful instrument made of what appeared to be mahogany with white enamel or ivory faces. As shown in Figure 12.1, it had ten scales on one side and eleven on the other. With it you could perform multiplication and division; calculate squares, square roots, cubes and cube roots; do trigonometric calculations; and do logarithmic calculations using logs to the base e or base 10. It was a marvel and it never needed batteries or an electric outlet. It was totally reliable and never gave you the message that, "This program has performed an illegal operation." All that was required to keep it from "locking up" or "freezing" was an occasional application of some wax on the sliding center section.

The slide rule was rendered obsolete, first by the scientific calculator and then by the computer. The speed and precision of these electronic marvels quickly displaced the slide rule, which at best could only give an answer to about three significant figures with any degree of accuracy at all. That wasn't a problem however because it was, and still is, sufficiently accurate for most engineering applications.

The slide rule did not tell you where the decimal point belonged. You had to figure that out for yourself. You had to know the expected magnitude of your answers but that generally was not a problem. If you knew what you were doing,

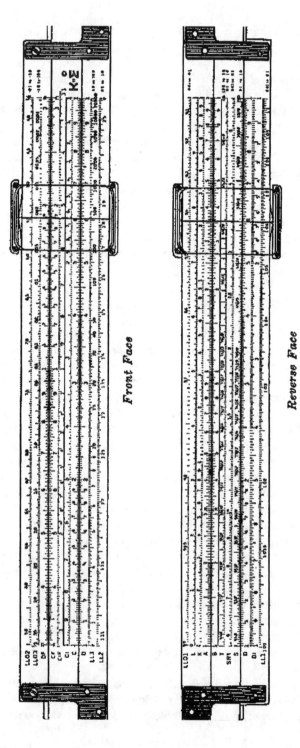

Front Face

Reverse Face

Figure 12.1 The Keuffel & Esser Log Log Duplex Decitrig slide rule.

you had a good sense of the magnitude of the correct answer. Today, computer answers are all too often accepted at face value.

Slide rule calculations took a lot more time than computer calculations do today. Complex designs were difficult and time consuming to perform. Consequently, many routine projects today would never have been attempted in the slide rule age. Computers are so fast that they can perform in a fraction of a second calculations that would have taken months or years with slide rules.

However, the computer cannot think. It is nothing more than a very fast, highly accurate machine that can do no more than human beings program into it. The computer will simply spit out an answer that would have been questioned by the slide rule engineer based upon his or her experience. The computer won't look at the answer quizzically and say, "Is this reasonable? Have I thought of everything?" Unfortunately, in the words of Duke University Professor Henry Petroski (1987), the computer "... gives us more information than we are able to assimilate. It gives the illusion of looking at everything. It also gives more apparent precision than can be meaningful ... I believe that we should never rely entirely on the computer to anticipate failure. In particular, we should not expect it to think of failure modes that human engineers and programmers haven't already thought of."

Professor Petroski goes on to say that, "Human engineers worry and lose sleep ..." thinking about the consequences of their designs. Computers don't worry. They just do what has been programmed into them—even if the programming has overlooked something, is incomplete, or is inadequate. Computers are fast, accurate, and can generate a false sense of security.

One classic example is the collapse of the computer-designed roof of the Hartford, Connecticut, Civic Center in 1978. This was a complex roof structure that probably would never have been attempted before the advent of computers. It was about 2-1/2 acres in area and was supported by only four columns. It provided spectators with an unobstructed view of the arena floor.

In January 1979, thousands of people attended a basketball game in that arena. It was a cold night and snow was falling. The roof also fell—fortunately after the game was over and the spectators had left. It was unable to support the weight of the accumulating snow and ice. If the snow had been coming down faster, the roof might have collapsed during the game, killing and injuring thousands of people.

The roof collapsed because the rods supporting it had buckled. This mode of failure was completely unanticipated by the computer. Professor Petroski commented, "The design would have been impossibly tedious for the slide rule engineer." The project probably would never have been undertaken except for the promised speed and reliability of the computer. Professor Petroski referred to this type of computer design problem as "... a new mode of failure, the computer-aided catastrophe."

The point of this discussion is that engineers, not computers, are responsible for the consequences of their designs. It is not sufficient to design by computer and proceed to build. The engineer has the ethical obligation to do everything possible to verify the suitability of the design for its intended purpose. Similarly, it is not sufficient to develop an estimate or project schedule using software without thoroughly understanding if the software is fully applicable to the project at hand. The first Canon of the *ICEC Canons of Ethics*, which is also the first Fundamental Canon of the NSPE *Code of Ethics*, must remain ever present in the engineer's mind. The engineer should never blindly accept a computer-generated design without question.

What constitutes professionalism? The most important part of the answer is the first Canon which says, that *"Engineers, in the fulfillment of their professional duties, shall hold paramount the safety, health, and welfare of the public."* Computers can't do that. Engineers must accept this responsibility personally.

12.5　INTERNATIONAL PROJECT ETHICS

A very pressing problem facing the engineering profession is how to work in a growing international business arena without compromising the codes of ethics of the profession. Engineering and construction firms are increasingly engaged in multinational projects and face many situations that are unlike what they are accustomed to in their home countries. Conditions are often different technically. Those can be readily handled. Ethical considerations, however, present even greater problems including:

- Inadequate environmental standards
- Sex discrimination
- Bribery to obtain favor with local officials
- Labor exploitation, in terms of wages, benefits, and safe working conditions
- Expected and expensive "gifts" to potential clients
- "Information brokering," i.e., selling of inside information to suppliers anxious to win contracts.

Some of these problems are so pervasive that the United Nations General Assembly adopted a treaty in October 2003 calling on governments to outlaw bribery, embezzlement and other corrupt practices. United Nations Secretary General Kofi Annan was quoted as saying, "The adoption of the United Nations Convention against Corruption sends a clear message that the international community is determined to prevent and control corruption ... and it reaffirms the importance of core values, such as honesty, respect for the rule of law, accountability and transparency, in promoting development and making the world a better place for all."

While the actions of the United Nations are a start toward eliminating some unethical practices, it would be naive to assume that they will stop. In particular, the practices of gift giving, information brokering, and bribery will continue so long as humans are motivated by greed. Nor will the other practices listed above end.

Engineering Times (June 1998) carried an editorial on international work that bears repeating. It said in part:

> Engineers involved in an international business situation that strains their professional code of ethics may wish that an interpreter of ethics abroad were at least as accessible as a translator of languages.
>
> In the current absence of a uniform international approach, a good sense of ethics may feel like a handicap to professionals trying to succeed within cultures that play by different rules. If engineers "do as the Romans do" in violation of their ethical code, they may make shortsighted economic gains, but in the long run, they can lose the respect of their peers and the international community—and break the law.
>
> Bribing public officials to influence the award of projects is just one example of an unethical practice encountered overseas. Outlawed by the Federal Corrupt Practices Act in the U. S., such situations nevertheless arise in numerous countries. Development of an international code of ethics that could be accepted by the world community would be an unequivocal step toward preventing such policies.

While the United Nations action is a step in the right direction, unfortunately there still is no international code of ethics that might help curtail those practices abroad which U.S. engineers see as being unethical. There is also no justification for "doing as the Romans do." Engineers must obey their codes of ethics, no matter what custom prevails at the project location. That may mean that the job will be lost in some cases but that is part of the price of practicing ethically.

Engineers who do think it is acceptable to use an agent abroad to "do the dirty work," thus acting like Pontius Pilate and "washing one's hands" of the problem also need to realize that they are committing a felony under U.S. law and are subject to prosecution.

At a recent seminar, a participant said that those who wish to engage in this practice have only one option, to give up their citizenship and become a citizen of the country in question. So long as they are U.S. citizens, the Federal Corrupt Practices Act applies to them. His comment also applies to corporations. A U.S. corporation is a citizen in the eyes of the law. Those companies engaging in bribery can avoid the law only by giving up their corporate charters and reincorporating abroad. Is your company willing to do that?

The message is clear. The only legal and ethical course of action is to forgo engaging in projects where bribes and gift giving are the accepted norm. Responsible firms will do just that and, in the long run, the entire engineering profession will be the benefactor.

REFERENCES

Ferenz, L. (1999). "Problem-Solving in Ethics," in *What Every Engineer Should Know About Ethics* (K. K. Humphreys), New York: Marcel Dekker, Inc., Chap. 1.

Petroski, H. (1987). *To Engineer is Human*. British Broadcasting Corporation television documentary. Distributed in the United States by Films Illustrated, a subsidiary of Public Media Films Inc.; distributed in Canada by BBC Television Distributors.

RECOMMENDED READING

Humphreys, K. K. (1999). *What Every Engineer Should Know About Ethics*. New York: Marcel Dekker, Inc.

Appendix A

AACE International Certification Programs

A.1 INTRODUCTION

Certification as a Certified Cost Engineer (CCE), Certified Cost Consultant (CCC), Planning and Scheduling Professional (PSP) or Interim Cost Consultant (ICC) requires passage of a written examination. The CCE and CCC certifications also require preparation and acceptance of a paper on a Cost Engineering subject. This appendix provides guidance for preparing to meet the requirements of these certification programs.

A.2 CERTIFIED COST ENGINEER/CERTIFIED COST CONSULTANT EXAMINATION

Exam Basis

The purpose of any professional certification or licensing program is to provide a mechanism to officially and publicly recognize the capabilities of an individual in a professional area. Certification as a Certified Cost Engineer or Certified Cost Consultant recognizes holders of those certificates as having capabilities encompassing the functions incorporated within the definition of Cost Engineering. Specifically:

> Cost Engineering is that area of engineering practice where engineering judgment and experience are utilized in the application of scientific principles and techniques to problems of cost estimation; cost control; business planning and management science; profitability analysis; and project management, planning and scheduling.

It is this definition of Cost Engineering that provides the basis for certification exam design—the exam must test proficiency across these areas.

275

To more specifically define cost engineering in terms of expected skills and knowledge, AACE International has published the *Required Skills and Knowledge of a Cost Engineer: AACE International Recommended Practice No. 11-R88*. But, even that document does not totally define Cost Engineering. The use of the word "required" in its title emphasizes that these are only baseline skills and knowledge; there are more advanced skills and knowledge to be found in practice. Further, Cost Engineering is a dynamic profession; it is affected by advances in philosophies, methodologies and technology. The true professional Cost Engineer is expected to keep abreast of these advances and to demonstrate this awareness in the exam.

In summary, the definition of Cost Engineering and the *Required Skills and Knowledge of a Cost Engineer* document are the basic "scope" documents for the certification exam.

Exam Schedule

There is a summer exam and a winter exam. The summer exam is always given in conjunction with the AACE International Annual Meeting at that meeting site; it may be given elsewhere at testing sites authorized by the Certification Board. The winter exam is commonly held in December at testing sites authorized by the Certification Board. Tests at sites other than that of the Annual Meeting are normally hosted by AACE International sections. Proctored examinations can also be arranged at other times and places. Consult AACE International for exact times and places.

Exam Format

The exam consists of four parts. You will have $1\frac{1}{2}$ hours for each part. Parts I and II will be given in the morning session with a 30 minute break between parts. Parts III and IV will be given during the afternoon session. Some Parts or portions of Parts will be Open Book; some Closed Book—the exam paper will specify in each case.

Proctors are assigned to administer the exams. They have no advance knowledge of exam content so cannot be expected to interpret exam questions for you.

The broad subject areas included within each part are listed below. Obviously, not all subjects are covered in each exam. New exams are prepared for each examination.

Part I.   Supporting Skills and Knowledge

        A.   Computer Operations/Operations Research
        B.   Measurements/Conversions/Statistics and Probability
        C.   Cost and Schedule Terminology/Basic Applications
        D.   Basic Business and Finance
        E.   Inflation

Part II. Cost Estimating and Control

    A. Elements of Cost
    B. Chart of Accounts/Work Breakdown Structure (WBS)
    C. Costing and Pricing
    D. Estimating Methods
    E. Types and Purpose of Estimates
    F. Operating/Manufacturing Costs
    G. Cost Indexes and Escalation Factors
    H. Risk Analysis/Contingency
    I. Budgeting and Cash Flow

Part III. Project Management

    A. Management Theory/Organizational Structures
    B. Behavioral Science/Motivational Management
    C. Integrated Project Control
    D. Planning and Scheduling
    E. Resource/Productivity Management
    F. Contracts and Contract Administration
    G. Social and Legal Issues in Management

Part IV. Economic Analysis

    A. Value Analysis/Value Engineering
    B. Depreciation
    C. Comparative Economic Studies
    D. Profitability
    E. Life Cycle Costs
    F. Time Value of Money/Engineering Economics
    G. Forecasting

Recognizing that there are many fields of interest within the profession—Engineering, Construction, Manufacturing, Process, Mining, Utilities, Transportation, Aerospace, Environmental and Government—you can expect questions in any of those settings. However, as a practical matter, no one is expected to be conversant in all areas and the exam is designed to take this into account through its multiple option format and extensive use of questions of general applicability.

The first section of the examination consists of 50 multiple choice questions worth 2 points each for a total of 100 points. Each of the other sections of the exam has a Part A and a Part B. Part A consists of questions that are worth 35 points. Part B consists of quick questions worth 1.5 points each. Two questions must be answered in Part A and twenty in Part B. Thus, an examinee has the potential for earning 100 points in each of the four exam sections for a total score of 400. As the scoring would indicate, Part A questions will more complex than Part B questions.

A group of representative questions from recent exams is included in this appendix.

Preparation Guidance

To assist in your preparation for the exam, you should first obtain a copy of the *Required Skills and Knowledge of a Cost Engineer* since it outlines much of what you need to know. From there you have a number of options:

In conjunction with the AACE International Annual Meeting, a *Fundamental Skills of Cost Engineering* seminar is held. This is one of the continuing education courses available at extra cost in conjunction with the annual meeting.

At most AACE International Annual Meetings there is a Skills and Knowledge track within the Technical Program on basic or advanced subjects in Cost Engineering. A typical track consists of 12–17 one-hour blocks of instruction. Attendance at any or all of these sessions is open to any meeting registrant. Specific subjects are listed on the Technical Program and described in its accompanying book of abstracts issued before each annual meeting.

A number of local sections of AACE International sponsor formal workshops on Cost Engineering, some specifically intended to assist candidates in exam preparation. If your section does not have such a workshop, start the ball rolling. Even if no formal workshop is available, consider a "bootstrap" style workshop wherein you and other candidates meet regularly to teach each other.

A correspondence course developed by the Department of Civil and Construction Engineering at Iowa State University is also available. For details, see the distance learning section of the AACE International web site (http://www.aacei.org) or contact Iowa State University, Extended & Continuing Education, 102 Scheman Building, Ames, IA 50011-1112, USA. Email: conted@exnet.iastate.edu.

AACE International also offers a four-day certification review course that incorporates one part of the examination at the end of each day of the course. This course is offered periodically at various locations and enables participants to immediately be examined rather than waiting until the next time the examinations are scheduled. Participants electing this option must complete an application for certification and submit it when registering for the course. The required professional paper must also be submitted prior to taking the written examination.

The Professional Paper

The professional paper to be submitted as part of the examination is to be of a quality suitable for presentation at an AACE International annual, regional or section meeting; or for publication in *Cost Engineering* magazine. The specific standards are detailed on the "Application for Examination as A Certified Cost

Engineer or Certified Cost Consultant" available from AACE International. You may be able to get credit for a recently published article for which you were primary author; see the application form for details.

Now for specific guidance:

A common mistake among inexperienced writers is to choose too broad a subject area. A paper of 2500 words (6–8 pages, single spaced) is not that long. So, choose a narrow topic area and give it good coverage.

If the paper involves a subject requiring employer review and approval, make certain your bosses know you intend to write on the subject and get their blessing up front so you won't waste time.

Unless you are an experienced writer, you will need to allow considerable time to develop your subject into a polished document. A good rule-of-thumb is to make a rough estimate of the time you think you will need and multiply by three.

Think about your paper when you're playing golf, traveling or otherwise have time just to think. You'll be surprised how ideas suddenly come to you. But, make certain you have some means to jot those ideas down before you forget them.

You should expect to go through several drafts before final. If you or your typist have access to a modern word processor, this process will be greatly simplified. After you complete each draft, set it aside for a couple days, then come back and review it to see if it makes sense and further refine it.

Keep a copy of the latest draft in your briefcase so you can review it when traveling or at other opportune times. Of course, it is always helpful to get an outside opinion or two on the paper both for content and presentation.

Most guide books on writing will tell you to outline your subject and then expand from the outline. This is fine, but often leads to "writer's block" since the writer feels compelled to write the abstract, introduction, main body, etc. in sequence and can't find the words to get started. Some authors, having faced this situation, have found that the best way to get going is to start the paper anywhere you feel comfortable, usually somewhere in the main body, and work from there. Initially, you don't even need to document your thoughts in any logical sequence—just write them down as they come to you. With modern word processors you can readily add, delete, and move text around later to provide orderly coverage. Using this approach, you will probably find that your paper's coverage evolves to be somewhat different than originally planned. This leads to the last bit of advice—wait until the main body is complete before writing your introduction and abstract. Had you written them first, they probably would no longer fit the paper anyway.

The paper can be given to proctors at the time of examination or can be mailed directly to AACE Headquarters. Two English language printed copies of the paper and an electronic text file must be submitted. Candidates may not sit for the examination until the paper is submitted.

Sample Exam Questions

Parts II–IV, Section A Example Questions
(Section A questions are each worth 35 points–5 points per multiple-choice answer)

A1.　In planning to acquire a major asset, you would need to investigate the cost implications and advantages/disadvantages of leasing or purchasing. Answer the following seven (7) questions about the relevant issues surrounding leases. (35 total points)

A1.1　All of the following are advantages of leasing except:
A.　No down payment required
B.　May provide favorable tax benefits
C.　Provides protection against equipment obsolescence
D.　Payments result in lower overall costs

A1.2　The most common type of lease is:
A.　Financial
B.　Operating
C.　Sale or leaseback
D.　Term

A1.3　All of the following are attributes of leasing except:
A.　Lessees often underestimate the salvage value of the asset
B.　Overall cost is higher because tax advantages of ownership are lost
C.　Lease agreement is usually a long term obligation
D.　Accelerated depreciation can be used

A1.4　The following statements about financial leases are true except:
A.　The lease is written for a term not to exceed the economic life of the asset
B.　The lessee is usually responsible for expenses such as maintenance, taxes, and insurance
C.　The lessee can always buy out of the lease before the end of the term
D.　The lease usually stipulates that periodic payments be made

A1.5　The following statements are true about operating lease attributes except:
A.　It is the primary form for the leasing of computer equipment
B.　It is the primary form for the leasing of motor vehicles

    C.   Maintenance of the asset is usually the responsibility of the lessee

    D.   Usually the lease can be canceled under conditions stipulated in the agreement

A1.6 Identify which statement about leasing is incorrect.

    A.   Requires some restrictions on a company's financial operations

    B.   The economic value of the asset is lost at the end of the leasing term

    C.   Discounted cash flow analysis is used to determine lease vs. buy decisions

    D.   None of the above

A1.7 The following statements are true about term leases except:

    A.   The term lease is generally used for lease periods greater than 5 years

    B.   Tax benefits are greatest under the term lease

    C.   The term lease can be depreciated

    D.   None of the above

A2.   It is necessary to perform operating and manufacturing cost estimates at both full plant capacity and at conditions other than full plant capacity. An analysis of costs at other than full plan capacity needs to take into account fixed, variable, and semivariable costs. Answer the following seven (7) questions relating to this scenario: (35 total points)

A2.1 An analysis of operating and maintenance costs at partial capacity enables the determination of:

    A.   Variable costs

    B.   Fixed costs

    C.   Semivariable costs

    D.   Breakeven point

A2.2 Those costs that are independent of the system throughput are:

    A.   Variable costs

    B.   Fixed costs

    C.   Operating costs

    D.   All of the above

A2.3 Semivariable costs could include:

    A.   Costs that are not directly fixed

    B.   Supervision

    C.   Plant overhead

    D.   All of the above

A2.4 Fixed costs include all of the following except:

    A.   Depreciation

    B.   Property taxes

    C.   Utilities

    D.   All of the above

A2.5 Another name for semivariable costs is:
   A. Proportional costs
   B. Production costs
   C. Manufacturing costs
   D. Packaging costs

A2.6 Royalty cost can be considered as:
   A. Variable costs
   B. Semivariable costs
   C. Fixed costs
   D. Any of the above

A2.7 An example of a fixed cost is:
   A. Depreciation
   B. General expense
   C. Plant overhead
   D. Material

Part I and Parts II–IV (CCC/CCE Exam–Section B example questions) (each worth 1.5 points)

B3.   As the purchasing manager for a local company located in the U.S., you receive a request from a purchasing manager representing a project buyer overseas for a price quotation for six (6) kilograms (kg) of babbitt for a project. Babbitt is priced in your firm for $3.62 per pound. What price would you quote (in U.S. dollars excluding freight) in reply to the request?
   A. $47.88   C. $4.78
   B. $21.72   D. $27.15

B4.   The following technique can be used to prepare an Order of Magnitude estimate:
   A. Parametric Estimate
   B. Monte Carlo Method
   C. Detailed Takeoffs and Pricing
   D. All of the above

B5.   Contract Management
      Which condition listed below would not enable Liquidated Damages to be applied to a contract?
   A. Delay in the competition of the overall contract
   B. Loss of revenue for the owner
   C. Acts of nature
   D. Unreasonable inconvenience

B6.   What are the three types of value?
   A. Use, Esteem, and Internal
   B. Use, Esteem, and Exchange

    C.   Esteem, Exchange, and Internal
    D.   Use, Exchange, and Internal

Answers to Sample Exam Questions

A1.1	D
A1.2	A
A1.3	D
A1.4	C
A1.5	B
A1.6	A
A1.7	D
A2.1	D
A2.2	B
A2.3	D
A2.4	C
A2.5	A
A2.6	D
A2.7	A
B3.	A. $47.88
B4.	A. Parametric Estimate
B5.	C. Acts of nature
B6.	B. Use, Esteem, and Exchange

A.3 INTERIM COST CONSULTANT EXAMINATION

Exam Basis, Schedule and Format

The Interim Cost Consultant program is specifically designed for young professionals looking for a way of obtaining outside, third party validation of their skills and knowledge in the cost and schedule areas but who lack sufficient practical experience to qualify to sit for the CCE/CCC examination. After attainment of eight or more years of experience, persons who have qualified under the ICC program may sit for the CCE/CCC examination. The topics of the ICC examination are the same as those for the CCE/CCC examination. Examination dates are normally the same as those for the CCE/CCE examination at the same testing locations. The examination is given in two 1 hour 45 minute parts consisting of 50 multiple choice questions per part.

ICC Sample Exam Questions

The questions on the ICC examination are the same types of questions that are asked in Part b of each section of the CCE/CCE examination. Sample questions of this type appear above.

A.4 PLANNING AND SCHEDULING PROFESSIONAL CERTIFICATION EXAMINATION

The PSP examination consists of four parts.

- Part I is Basic Knowledge. It consists of multiple-choice questions concerning the basics of planning and scheduling.
- Part II is Planning and Scheduling Applications. It consists of multiple-choice questions involving planning and scheduling scenarios.
- Part III is a Practical Exercise. This part entails solving one to three problems and answering a series of multiple-choice questions concerning the problem(s).
- Part IV is a real-time communications exercise. It requires the candidate drafting a one page (maximum) memorandum to simulate reporting a planning and scheduling analysis to the project manager explaining the issues and proposing solution(s) regarding a given problem scenario.

If you should have any questions regarding the AACE International Certification Examinations or Certification procedures, please contact the Certification Secretary at AACE International Headquarters, 209 Prairie Ave., Suite 100, Morgantown, WV 26501, USA. Phone: 1-800-858-COST or 1-304-296-8444. E-mail: info@aacei.org.

Appendix B

Standard Cost Engineering Terminology*

Account code structure—the system used to assign summary numbers to elements of the work breakdown and account numbers to individual work packages.

Account number—a numeric identification of a work package. An account number may be assigned to one or more activities. Syn.: Shop Order Number.

Accounts payable—the value of goods and services rendered on which payment has not yet been made.

Accounts receivable—the value of goods shipped or services rendered to a customer on which payment has not yet been received. Usually includes an allowance for bad debts.

Activity—a basic element of work, or task that must be performed in order to complete a project. An activity occurs over a given period of time.

Activity code—any combination of letters, numbers, or blanks that describes and identifies any activity or task shown on the schedule. Syn.: Activity Identifier.

Activity splitting—dividing (i.e., splitting) an activity of stated scope, description and schedule into two or more activities that are rescoped and rescheduled. The sum of the split activities is normally the total of the original.

Activity total slack—the latest allowable end time minus earliest allowable end time. The activity slack is always greater than or equal to the slack of the activity ending event.

---

*This Appendix is an abridgement of AACE Standard No. 10S-90 as revised through January 2003. For the complete standard, or for official definitions of terms not included here, refer to the full standard, published in the *AACE International Recommended Practices and Standards*, AACE International, 209 Prairie Avenue, Suite 100, Morgantown, West Virginia 26501, USA.

Actual cost of work performed (ACWP)—the direct costs actually incurred and the direct costs actually recorded and assigned in accomplishing the work performed. These costs should reconcile with the contractor's incurred cost ledgers when they are audited by the client.

Administrative expense—the overhead cost due to the nonprofit-specific operations of a company. Generally includes top management salaries and the costs of legal, central purchasing, traffic, accounting, and other staff functions and their expenses for travel and accommodations.

Allowances—additional resources included in estimates to cover the cost of known but undefined requirements for an individual activity, work item, account or subaccount.

Amortization—(1) as applied to a capitalized asset, the distribution of the initial cost by periodic charges to operations as in depreciation. Most properly applies to assets with indefinite life; (2) the reduction of a debt by either periodic or irregular payments; (3) a plan to pay off a financial obligation according to some prearranged schedule.

Annuity—(1) an amount of money payable to a beneficiary at regular intervals for a prescribed period of time out of a fund reserved for that purpose; (2) a series of equal payments occurring at equal periods of time.

Arrow—the graphic representation of an activity in the ADM network. One arrow represents one activity. The tail (see I-NODE) of the arrow represents the start of the activity. The head (see J-NODE) of the arrow represents the finish. The arrow is not a vector quantity and is not drawn to scale. It is uniquely defined by two event codes.

Arrow diagramming method (ADM)—a method of constructing a logical network of activities using arrows to represent the activities and connecting them head to tail. This diagramming method shows the sequence, predecessor and successor relationships of the activities.

Average-interest method—a method of computing required return on investment based on the average book value of the asset during its life or during a specified study period.

Backward pass—calculation of the latest finish time and latest start time for all uncompleted network activities or late time for events in the ADM and PDM methods. It is determined by working from the final activity and subtracting durations from uncompleted activities.

Bar chart—a graphic presentation of project activities shown by a time-scaled bar line. Syn.: Gantt Chart.

Base period (of a given price index)—period for which prices serve as a reference for current period prices; in other words, the period for which an index is defined as 100 (if expressed in percentage form) or as 1 (if expressed in ratio form).

Battery limit—comprises one or more geographic boundaries, imaginary or real, enclosing a plant or unit being engineered and/or erected, established for the purpose of providing a means of specifically identifying certain portions of the plant, related groups of equipment, or associated facilities. It generally refers to the processing area and includes all the process equipment, and excludes such other facilities as storage, utilities, administration buildings, or auxiliary facilities. The scope included within a battery limit must be well-defined so that all personnel will clearly understand it.

Beginning event—an event that signifies the beginning of an activity. Syn.: Predecessor Event; Preceding Event; Starting Event.

Beginning (start) node of network—(ADM) a node at which no activities end, but one or more activities begin.

Benchmarking—A measurement and analysis process that compares practices, processes, and relevant measures to those of a selected basis of comparison (i.e., the benchmark) with the goal of improving performance. The comparison basis includes internal or external competitive or best practices, processes or measures. Examples of measures include estimated costs, actual costs, schedule durations, resource quantities and so on.

Block diagram—a diagram made up of vertically placed rectangles situated adjacent to each other on a common base line. Where the characteristic to be depicted is quantitative, the height of the rectangles is usually taken to be proportional to this quantitative variable. When this kind of diagram is used to portray a frequency distribution it takes the name of histogram.

Book value (net)—(1) current investment value on the books calculated as original value less depreciated accruals; (2) new asset value for accounting use; (3) the value of an outstanding share of stock of a corporation at any one time, determined by the number of shares of that class outstanding.

Breakeven point—(1) in business operations, the rate of operations output, or sales at which income is sufficient to equal operating costs or operating cost plus additional obligations that may be specified; (2) the operating condition, such as output, at which two alternatives are equal in economy; (3) the percentage of capacity operation of a manufacturing plant at which income will just cover expenses.

Breakout schedule—this jobsite schedule, generally in bar chart form is used to communicate the day-to-day activities to all working levels on the project as directed by the construction manager. Detail information with regard to equipment use, bulk material requirements, and craft skills distribution, as well as the work to be accomplished, forms the content of this schedule. The schedule is issued on a weekly basis with a two- to three-week look ahead from the issue date. This schedule generally contains from 25 to 100 activities.

Budget—a planned allocation of resources. The planned cost of needed materials is usually subdivided into quantity required and unit cost. The planned cost of labor is usually subdivided into the workhours required and the wage rate (plus fringe benefits and taxes).

Budget cost of work performed (BCWP)—the sum of the budgets for completed portions of in-process work, plus the appropriate portion of the budget for level of effort and apportioned effort for the relevant time period BCWP is commonly referred to as "earned value."

Budget cost of work scheduled (BCWS)—the sum of the budgets for work scheduled to be accomplished (including work-in-process), plus the appropriate portion of the budgets for level of effort and apportioned effort for the relevant time period.

Budgeting—A process used to allocate the estimated cost of resources into cost accounts (i.e., the cost budget) against which cost performance will be measured and assessed. Budgeting often considers time-phasing in relation to a schedule and/or time-based financial requirements and constraints.

Bulk material—material bought in lots. These items can be purchased from a standard catalog description and are bought in quantity for distribution as required. Examples are pipe (nonspooled), conduit, fittings, and wire.

Burden—in construction, the cost of maintaining an office with staff other than operating personnel. Includes also federal, state, and local taxes, fringe benefits and other union contract obligations. In manufacturing, burden sometimes denotes overhead.

Capacity factor—(1) the ratio of average load to maximum capacity; (2) the ratio between average load and the rated capacity of the apparatus; (3) the ratio of the average actual use to the rated available capacity. Also called Capacity Utilization Factor.

Capital budgeting—a systematic procedure for classifying, evaluating, and ranking proposed capital expenditures for the purpose of comparison and selection, combined with the analysis of the financing requirements.

Capital, direct—see Direct cost (1).

Capital, fixed—the total original value of physical facilities that are not carried as a current expense on the books of account and for which depreciation is allowed by the Federal Government. It includes plant equipment, building, furniture and fixtures, transportation equipment used directly in the production of a product or service. It includes all costs incident to getting the property in place and in operating condition, including legal costs, purchased patents, and paid-up licenses. Land, which is not depreciable, is often included. Characteristically it cannot be converted readily into cash.

Capital, indirect—see Indirect cost (1).

Capital, operating—capital associated with process facilities inside battery limits.

Capital recovery—(1) charging periodically to operations amounts that will ultimately equal the amount of capital expenditure (see Amortization, Depletion, and Depreciation); (2) the replacement of the original cost of an asset plus interest; (3) the process of regaining the net investment in a project by means of revenue in excess of the costs from the project.

(Usually implies amortization of principal plus interest on the diminishing unrecovered balance.)

Capital recovery factor—a factor used to calculate the sum of money required at the end of each of a series of periods to regain the net investment of a project plus the compounded interest on the unrecovered balance.

Capital, venture—capital invested in technology or markets new at least to the particular organization.

Capital, working—the funds in addition to fixed capital and land investment that a company must contribute to the project (excluding startup expense) to get the project started and meet subsequent obligations as they come due. Working capital includes inventories, cash and accounts receivable minus accounts payable. Characteristically, these funds can be converted readily into cash. Working capital is normally assumed recovered at the end of the project.

Capitalized cost—(1) the present worth of a uniform series of periodic costs that continue for an indefinitely long time (hypothetically infinite); (2) the value at the purchase date of the asset of all expenditures to be made in reference to this asset over an indefinite period of time. This cost can also be regarded as the sum of capital that, if invested in a fund earning a stipulated interest rate, will be sufficient to provide for all payments required to maintain the asset in perpetual service.

Cash costs—total cost excluding capital and depreciation spent on a regular basis over a period of time, usually one year. Cash costs consist of manufacturing cost and other expenses such as transportation cost, selling expense, research and development cost or corporate administrative expense.

Cash flow—the net flow of dollars into or out of a project. The algebraic sum, in any time period, of all cash receipts, expenses, and investments. Also called cash proceeds or cash generated.

Chebyshev's theorem—a statistical method of predicting the probability that a value will occur within one or more standard deviations ($\pm$) of the mean.

Code of accounts (COA)—A systematic coding structure for organizing and managing asset, cost, resource, and schedule activity information. A COA is essentially an index to facilitate finding, sorting, compiling, summarizing, and otherwise managing information that the code is tied to. A complete code of accounts includes definitions of the content of each account. Syns.: Chart of Accounts, Cost Codes.

Commodity—in price index nomenclature, a good and sometimes a service.

Composite price index—an index that globally measures the price change of a range of commodities.

Compound amount—the future worth of a sum invested (or loaned) at compound interest.

Compound amount factor—(1) the function of interest rate and time that determines the compound amount from a stated initial sum; (2) a factor which when multiplied by the single sum or uniform series of payments will give the future worth at compound interest of such single sum or series.

Compound interest—(1) the type of interest to which is periodically added the amount of investment (or loan) so that subsequent interest is based on the cumulative amount; (2) the interest charges under the condition that interest is charged on any previous interest earned in any time period, as well as on the principal.

Compounding, continuous—(1) a compound interest situation in which the compounding period is zero and the number of periods infinitely great. A mathematical concept that is practical for dealing with frequent compounding and small interest rates; (2) a mathematical procedure for evaluating compound interest factors based on a continuous interest function rather than discrete interest periods.

Compounding period—the time interval between dates at which interest is paid and added to the amount of an investment or loan. Designates frequency of compounding.

Constant utility price index—a composite price index which measures price changes by comparing the expenditures necessary to provide substantially equivalent sets of goods and services at different points in time.

Construction cost—the sum of all costs, direct and indirect, inherent in converting a design plan for material and equipment into a project ready for start-up, but not necessarily in production operation; the sum of field labor, supervision, administration, tools, field office expense, materials, and equipment.

Consumer price index (CPI)—a measure of time-to-time fluctuations in the price of a quantitatively constant market basket of goods and services, selected as representative of a special level of living.

Contingency—An amount added to an estimate to allow for items, conditions, or events for which the state, occurrence, and/or effect is uncertain and that experience shows will likely result, in aggregate, in additional costs. Typically estimated using statistical analysis or judgment based on past asset or project experience. Contingency usually excludes; 1) major scope changes such as changes in end product specification, capacities, building sizes, and location of the asset or project (see management reserve), 2) extraordinary events such as major strikes and natural disasters, 3) management reserves, and 4) escalation and currency effects. Some of the items, conditions, or events for which the state, occurrence, and/or effect is uncertain include, but are not limited to, planning and estimating errors and omissions, minor price fluctuations (other than general escalation), design developments and changes within the scope, and variations in market and environmental conditions. Contingency is generally included in most estimates, and is expected to be expended to some extent.

Contracts—legal agreements between two or more parties, which may be of the types enumerated below:

1. In Cost Plus contracts the contractor agrees to furnish to the client services and material at actual cost, plus an agreed upon fee for these services. This type of contract is employed most often when the scope of services to be provided is not well defined.

   a. Cost Plus Percentage Burden and Fee—the client will pay all costs as defined in the terms of the contract, plus "burden and fee" at a specified percent of the labor costs that the client is paying for directly. This type of contract generally is used for engineering services. In contracts with some governmental agencies, burden items are included in indirect cost.

   b. Cost Plus Fixed Fee—the client pays costs as defined in the contract document. Burden on reimbursable technical labor cost is considered in this case as part of cost. In addition to the costs and burden, the client also pays a fixed amount as the contractor's "fee."

   c. Cost Plus Fixed Sum—the client will pay costs defined by contract plus a fixed sum which will cover "non-reimbursable" costs and provide for a fee. This type of contract is used in lieu of a cost plus fixed fee contract where the client wishes to have the contractor assume some of the risk for items which would be Reimbursable under a Cost Plus Fixed Fee type of contract.

   d. Cost Plus Percentage Fee—the client pays all costs, plus a percentage for the use of the contractor's organization.

2. Fixed Price types of contract are ones wherein a contractor agrees to furnish services and material at a specified price, possibly with a mutually agreed upon escalation clause. This type of contract is most often employed when the scope of services to be provided is well defined.

   a. Lump Sum—contractor agrees to perform all services as specified by the contract for a fixed amount. A variation of this type may include a turnkey arrangement where the contractor guarantees quality, quantity and yield on a process plant or other installation.

   b. Unit Price—contractor will be paid at an agreed upon unit rate for services performed. For example, technical work-hours will be paid for at the unit price agreed upon. Often field work is assigned to a subcontractor by the prime contractor on a unit price basis.

   c. Guaranteed Maximum (Target Price)—a contractor agrees to perform all services as defined in the contract document guaranteeing that the total cost to the client will not exceed a stipulated maximum figure. Quite often, these types of contracts will contain special share-of-the-saving arrangements to provide incentive to the contractor to minimize costs below the stipulated maximum.

    d. Bonus-Penalty—a special contractual arrangement usually between a client and a contractor wherein the contractor is guaranteed a bonus, usually a fixed sum of money, for each day the project is completed ahead of a specified schedule and/or below a specified cost, and agrees to pay a similar penalty for each day of completion after the schedule date or over a specified cost up to a specified maximum either way. The penalty situation is sometimes referred to as liquidated damages.

Cost—in project control and accounting, it is the amount measured in money, cash expended or liability incurred, in consideration of goods and/or services received. From a total cost management perspective, cost may include any investment of resources in strategic assets including time, monetary, human, and physical resources.

Cost accounting—The historical reporting of actual and/or committed disbursements (costs and expenditures) on a project. Costs are denoted and segregated within cost codes that are defined in a chart of accounts. In project control practice, cost accounting provides the measure of cost commitment and/or expenditure that can be compared to the measure of physical completion (or earned value) of an account.

Cost analysis—a historical and/or predictive method of ascertaining for what purpose expenditures on a project were made and utilizing this information to project the cost of a project as well as costs of future projects. The analysis may also include application of escalation, cost differentials between various localities, types of buildings, types of projects, and time of year.

Cost approach—one of the three approaches in the appraisal process. Underlying the theory of the cost approach is the principle of substitution, which suggests that no rational person will pay more for a property than the amount with which he/she can obtain, by purchase of a site and construction of a building without undue delay, a property of equal desirability and utility.

Cost and schedule control systems criteria (C/SCSC)—established characteristics that a contractor's internal management control system must possess to assure effective planning and control of contract work, costs, and schedules.

Cost control—the application of procedures to monitor expenditures and performance against progress of projects or manufacturing operations; to measure variance from authorized budgets and allow effective action to be taken to achieve minimum costs.

Cost engineer—an engineer whose judgement and experience are utilized in the application of scientific principles and techniques to problems of estimation;

cost control; business planning and management science; profitability analysis; and project management, planning and scheduling.

Cost estimating—A predictive process used to quantify, cost, and price the resources required by the scope of an asset investment option, activity, or project. As a predictive process, estimating must address risks and uncertainties. The outputs of estimating are used primarily as inputs for budgeting, cost or value analysis, decision making in business, asset and project planning, or for project cost and schedule control processes.

As applied in the project engineering and construction industry, cost estimating is the determination of quantity and the predicting and forecasting, within a defined scope, of the costs required to construct and equip a facility. Costs are determined utilizing experience and calculating and forecasting the future cost of resources, methods, and management within a scheduled time frame. Included in these costs are assessments and an evaluation of risks.

Cost estimating relationship (CER)—In estimating, an algorithm or formula that is used to perform the costing operation. CERs show some resource (e.g., cost, quantity, or time) as a function of one or more parameters that quantify scope, execution strategies, or other defining elements. A CER may be formulated in a manner that in addition to providing the most likely resource value, also provides a probability distribution for the resource value. Cost estimating relationships may be used in either definitive or parametric estimating methods. See Definitive estimate and Parametric estimate.

Cost index (price index)—a number that relates the cost of an item at a specific time to the corresponding cost at some arbitrarily specified time in the past. See Price index.

Costing—Cost estimating activity that translates quantified technical and programmatic scope information into expressions of the cost and resources required. In costing, this translation is usually done using algorithms or CERs. In the cost estimating process, costing follows scope determination and quantification, and precedes pricing and budgeting.

Costing, activity based (ABC)—Costing in a way that the costs budgeted to an account truly represent all the resources consumed by the activity or item represented in the account.

Cost of capital—A term, usually used in capital budgeting, to express as an interest rate percentage the overall estimated cost of investment capital at a given point in time, including both equity and borrowed funds.

Cost of ownership—the cost of operations, maintenance, follow-on logistical support, and end item and associated support systems. Syn.: Operating and Support Costs.

Cost of quality—consists of the sum of those costs associated with: (a) cost of quality conformance, (b) cost of quality nonconformance, (c) cost of lost business advantage.

CPM—See Critical path method.

Critical activity—any activity on a critical path.

Critical path—sequence of jobs or activities in a network analysis project such that the total duration equals the sum of the durations of the individual jobs in the sequence. There is no time leeway or slack (float) in activity along critical path (i.e., if the time to complete one or more jobs in the critical path increases, the total production time increases). It is the longest time path through the network.

Critical path method (CPM)—a scheduling technique using arrow, precedence, or PERT diagrams to determine the length of a project and to identify the activities and constraints on the critical path.

Current cost accounting (CCA)—a methodology prescribed by the Financial Accounting Board to compute and report financial activities in constant dollars.

Declining balance depreciation—method of computing depreciation in which the annual charge is a fixed percentage of the depreciated book value at the beginning of the year to which the depreciation applies. Syn.: Percent on Diminishing Value.

Definitive estimate—In estimating practice, describes estimating algorithms or cost estimating relationships that are not highly probabilistic in nature (i.e., the parameters or quantification inputs to the algorithm tend to be conclusive or definitive representations of the scope). Typical definitive estimate algorithms include, but are not limited to, detailed unit and line-item cost techniques (i.e., each specific quantified item is listed and costed separately).

Deflation—an absolute price decline for a commodity; also, an operation by means of which a current dollar value series is transformed into a constant dollar value series (i.e., is expressed in "real" terms using appropriate price indexes as deflators).

Deliverable—a report or product of one or more tasks that satisfy one or more objectives and must be delivered to satisfy contractual requirements.

Demand factor—(1) the ratio of the maximum instantaneous production rate to the production rate for which the equipment was designed; (2) the ratio between the maximum power demand and the total connected load of the system.

Demurrage—a charge made on cars, vehicles, or vessels held by or for consignor or consignee for loading or unloading, for forwarding directions or for any other purpose.

Depletion—(1) a form of capital recovery applicable to extractive property (e.g., mines). Depletion can be on a unit-of-output basis related to original or current appraisal of extent and value of the deposit. (Known as percentage depletion.) (2) lessening of the value of an asset due to a decrease in the quantity available. Depletion is similar to depreciation except that it refers to such natural resources as coal, oil, and timber in forests.

Depreciated book value—the first cost of the capitalized asset minus the accumulation of annual depreciation cost charges.

Depreciation—(1) decline in value of a capitalized asset; (2) a form of capital recovery applicable to a property with a life span of more than one year, in which an appropriate portion of the asset's value is periodically charged to current operations.

Deterministic model—a deterministic model, as opposed to a Stochastic model, is one which contains no random elements and for which, therefore, the future course of the system is determined by its state at present (and/or in the past).

Development costs—those costs specific to a project, either capital or expense items, that occur prior to commercial sales and are necessary in determining the potential of that project for consideration and eventual promotion. Major cost areas include process, product, and market research and development.

Deviation costs—the sum of those costs, including consequential costs such as schedule impact, associated with the rejection or rework of a product, process, or service due to a departure from established requirements. Also may include the cost associated with the provision of deliverables that are more than required.

Direct cost—(1) in construction, cost of installed equipment, material and labor directly involved in the physical construction of the permanent facility. (2) in manufacturing, service and other nonconstruction industries, the portion of operating costs that is generally assignable to a specific product or process area. Usually included are:

    a. Input materials
    b. Operating, supervision, and clerical payroll
    c. Fringe benefits
    d. Maintenance
    e. Utilities
    f. Catalysts, chemicals and operating supplies
    g. Miscellaneous (royalties, services, packaging, etc.)

Definitions of the above classifications are:

    a. Input Material—raw materials that appear in some form as a product. For example, water added to resin formulation is an input material, but sulfuric acid catalyst, consumed in manufacturing high octane alkylate, is not.
    b. Operating, Supervision, and Clerical Payroll—wages and salaries paid to personnel who operate the production facilities.
    c. Fringe Benefits—payroll costs other than wages not paid directly to the employee. They include costs for:
        1. Holidays, vacations, sick leave

    2.   Federal old age insurance

    3.   Pensions, life insurance, savings plans, etc.

    In contracts with some governmental agencies these items are included in indirect cost.

d.   Maintenance Cost—expense incurred to keep manufacturing facilities operational. It consists of:

    1.   Maintenance payroll cost

    2.   Maintenance materials and supplies cost

    Maintenance materials that have a life of more than one year are usually considered capital investment in detailed cash flow accounting.

e.   Utilities—the fuel, steam, air, power and water that must be purchased or generated to support the plant operation.

f.   Catalysts, Chemicals and Operating Supplies—materials consumed in the manufacturing operation, but not appearing as a product. Operating supplies are a minor cost in process industries and are sometimes assumed to be in the maintenance materials estimate; but in many industries, mining for example, they are a significant proportion of direct cost.

g.   Miscellaneous

    1.   Costs paid to others for the use of a proprietary process. Both paid-up and "running" royalties are used. Cost of paid-up royalties are usually on the basis of production rate. Royalties vary widely, however, and are specific for the situation under consideration.

    2.   Packaging Cost—material and labor necessary to place the product in a suitable container for shipment. Also called Packaging and Container Cost or Packing Cost. Sometimes considered an indirect cost together with distribution costs such as for warehousing, loading and transportation.

    3.   Although the direct costs described above are typical and in general use, each industry has unique costs that fall into the "direct cost" category. A few examples are equipment rental, waste disposal, contracts, etc.

Discounted cash flow—(1) the present worth of a sequence in time of sums of money when the sequence is considered as a flow of cash into and/or out of an economic unit; (2) an investment analysis that compares the present worth of projected receipts and disbursements occurring at designated future times in order to estimate the rate of return from the investment or project. Also called Discounted Cash Flow Rate of Return, Interest Rate of Return, Internal Rate of Return, Investor's Method or Profitability Index.

Distribution curve—the graph of cumulated frequency as ordinate against the variate value as abscissa, namely the graph of the distribution function. The curve is sometimes known as an "ogive", a name introduced by

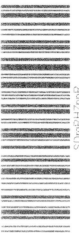

SDrgRHZppR

At www.Target.com/return, you can view our Return Policy and return options. We'll tell you whether your item can be returned in store (by printing a receipt) or by mail (by printing a pre-paid label).

NOTE: This packing slip cannot be used to return items to a Target store.

Your order of June 21, 2008 (Order ID 602 – 7356487 – 6409867)

To: JTTH
To This Shipment

1 **Project and Cost Engineers' Handbook, Fourth Edition (Cost Engineering)**
Paperback ISBN 0824757467
0824757467
Sold by Amazon.com/Target.com Inc. (Hardcover)

2 **Handbook for Process Plant Project Engineers**
Hardcover ISBN 1860583709
1860583709
Sold by Amazon.com/Target.com Inc. (Hardcover)

This shipment completes your order.

3627 (2 of 2)

Galton, because the distribution curve of a normal function is of the ogive shape; but not all distribution curves have this form and the term "ogive" is better avoided or confined to the normal or nearly normal case.

Dummy activity—an activity, always of zero duration, used to show logical dependency when an activity cannot start before another is complete, but which does not lie on the same path through the network. Normally, these dummy activities are graphically represented as a dashed line headed by an arrow and inserted between two nodes to indicate a precedence relationship or to maintain a unique numbering of concurrent activities.

Dummy start activity—an activity entered into the network for the sole purpose of creating a single start for the network.

Durable goods—generally, any producer or consumer goods whose continuous serviceability is likely to exceed three years (e.g., trucks, furniture).

Dynamic programming—a method for optimizing a set of decisions that may be made sequentially. Characteristically, each decision may be made in the light of the information embodied in a small number of observables called state variables. The incurred cost for each period is a mathematical function of the current state and decision variables, while future states are functions of these variables. The aim of the decision policy is to minimize the total incurred cost, or equivalently the average cost per period. The mathematical treatment of such decision problems involves the theory of functional equations, and usually requires a digital computer for implementation.

Early event time (EV)—the earliest time at which an event may occur.

Early finish time (EF)—the earliest time at which an activity can be completed; equal to the early start of the activity plus its remaining duration.

Early start time (ES)—the earliest time any activity may begin as logically constrained by the network for a specific work schedule.

Early work schedule—predicated on the parameters established by the proposal schedule and any negotiated changes, the early work schedule defines reportable pieces of work within major areas. The format is developed into a logic network including engineering drawings, bid inquiries, purchase orders, and equipment deliveries, and can be displayed as a time-phased network. The detail of this schedule concentrates on projected engineering construction issue drawings released and equipment deliveries. The activities of the early part of construction are more defined than in the proposal or milestone schedule.

Earned value—the periodic, consistent measurement of work performed in terms of the budget planned for that work. In criteria terminology, earned value is the budgeted cost of work performed. It is compared to the budgeted cost of work scheduled (planned) to obtain schedule performance and it is compared to the actual cost of work performed to obtain cost performance.

Earned value concept—the measurement at any time of work accomplished (performed) in terms of budgets planned for that work, and the use of these data to indicate contract cost and schedule performance. The earned value of work done is quantified as the budgeted cost for work performed (BCWP) compared to the budgeted cost for work scheduled (BCWS) to show schedule performance and compared to the actual cost of work performed (ACWP) to indicate cost performance.

Earned value reports—cost and schedule performance reports that are part of the performance measurement system. These reports make use of the earned value concept of measuring work accomplishment.

Earnings value—the present worth of an income producer's probable future net earnings, as prognosticated on the basis of recent and present expense and earnings and the business outlook.

Economic life—that period of time over which an investment is considered to be the least-cost alternative for meeting a particular objective. Syn.: Project Life.

Effective interest—the true value of interest rate computed by equations for compound interest rate for a 1-year period.

Ending event—the event that signifies the completion of all activities leading to that event.

Ending node of network (ADM)—a node where no activities begin, but one or more activities end.

End network event—the event that signifies the end of a network.

Equivalent sets of commodities—sets of commodities that provide the same total satisfaction to a given group of consumers (without necessarily being identical).

Escalation—the provision in actual or estimated costs for an increase in the cost of equipment, material, labor, etc, over that specified in the purchase order or contract due to continuing price level changes over time.

Escalator clause—clause contained in collective agreements, providing for an automatic price adjustment based on changes in specified indices.

Estimate, cost—an evaluation of all the costs of the elements of a project or effort as defined by an agreed-upon scope. Three specific types based on degree of definition of a process industry plant are:

1. Order of Magnitude Estimate—an estimate made without detailed engineering data. Some examples would be: an estimate from cost capacity curves, an estimate using scaleup or down factors, and an approximate ratio estimate. It is normally expected that an estimate of this type would be accurate within plus 50 percent or minus 30 percent.
2. Budget Estimate—budget in this case applies to the owner's budget and not to the budget as a project control document. A budget estimate is prepared with the use of flow sheets, layouts and equipment details. It is normally expected that an estimate of this type would be accurate within plus 30 percent or minus 15 percent.

3. Definitive Estimate—as the name implies, this is an estimate prepared
    from very defined engineering data. The engineering data includes as a
    minimum, nearly complete plot plans and elevations, piping and
    instrument diagrams, one line electrical diagrams, equipment data
    sheets and quotations, structural sketches, soil data and sketches of
    major foundations, building sketches and a complete set of specifica-
    tions. This category of estimate covers all types from the minimum
    described above to the maximum definitive type that would be made
    from "approved for construction" drawings and specifications. It is
    normally expected that an estimate of this type would be accurate
    within plus 15 percent and minus 5 percent.

Event—an identifiable single point in time on a project. Graphically, it is represented by a node. An event occurs only when all work preceding it has
been completed. It has zero duration.

Event name—an alphanumeric description of an event.

Event number—a numerical description of an event for computation and
identification.

Event slack—the difference between the latest allowable date and the earliest
date for an event.

Event times—time information generated through the network analysis calculation, which identifies the start and finish times for each event in the
network.

Exempt—employees exempt from federal wage and hours guidelines.

Expected begin date—begin date assigned to a specific activity. Syn.: Target
Start Date.

Expense—expenditures of short-term value, including depreciation, as
opposed to land and other fixed capital. For factory expense, see Plant
overhead.

Extrapolation—to infer from values within an observed interval, or to project or
extend beyond observed data.

Factor analysis—a branch of multivariate analysis in which the observed variates x_i $(i = 1, 2 \ldots, p)$ are supposed to be expressible in terms of a
number $m < p$ factors f_j, together with residual elements.

Feedback—information (data) extracted from a process or situation and used in
controlling (directly) or in planning or modifying immediate or future
inputs (actions or decisions) into the process or situation.

Feedback loop—the part of a closed-loop system that allows the comparison of a
response to a command.

Field cost—engineering and construction costs associated with the construction
site rather than with the home office.

Field labor overhead—the sum of the cost of payroll burden, temporary construction facilities, consumables, field supervision, and construction tools
and equipment.

Field supervision—the cost of salaries and wages of all field supervisory and field support staff personnel (excluding general foreman), plus associated payroll burdens, home office overhead, living and travel allowances, and field office operating costs.

FIFO (First In, First Out)—a method of determining the cost of inventory used in a product. In this method, the costs of materials are transferred to the product in chronological order. Also used to describe the movement of materials (see LIFO).

Fixed cost—those costs independent of short term variations in output of the system under consideration. Includes such costs as maintenance; plant overhead; and administrative, selling and research expense. For the purpose of cash flow calculation, depreciation is excluded (except in income tax calculations).

Float—(1) in manufacturing, the amount of material in a system or process, at a given point in time, that is not being directly employed or worked upon. (2) in construction, the cushion or slack in any noncritical path in a network planning system. Syn.: Slack, Path Float.

Forward pass—(1) in construction, network calculations that determine the earliest start/earliest finish time (date) of each activity. (2) in manufacturing, often referred to as forward scheduling, a scheduling technique where the scheduler proceeds from a known start date and computes the completion date for an order usually proceeding from the first operation to the last.

Free float (FF)—the amount of time that the completion of an activity may exceed its scheduled finish time without increasing the start time of any succeeding activity.

Frequency distribution—a specification of the way in which the frequencies of members of a population are distributed according to the values of the variates which they exhibit. For observed data the distribution is usually specified in tabular form, with some grouping for continuous variates. A conceptual distribution is usually specified by a frequency function or a distribution function.

Frequency function—an expression giving the frequency of a variate value x as a function of x; or, for continuous variates, the frequency in an elemental range dx. Unless the contrary is specified, the total frequency is taken to be unity, so that the frequency function represents the proportion of variate values x. From a more sophisticated standpoint the frequency function is most conveniently regarded as the derivative of the Distribution Function. The derivative is also commonly called the probability density function. The generalization to more than one variate is immediate.

Fringe benefits—employee welfare benefits, i.e., expenses of employment such as holidays, sick leave, health and welfare benefits, retirement fund, training, supplemental union benefits, etc.

Functional replacement cost—the current cost of acquiring the same service potential as embodied by the asset under consideration.

Functional worth—the lowest overall cost for performing a function. Four types are as follows:

Cost Value—the monetary sum of labor, material, burden, and all other elements of cost required to produce an item or provide a service.

Esteem Value—the monetary measure of the properties of a product or service, that contribute to desirability or salability but not to required functional performance.

Exchange Value—the monetary sum at which a product or service can be traded.

Use Value—the monetary measure of the necessary functional properties of a product or service that contribute to performance.

Gantt chart—see Bar chart.

General overhead—the fixed cost in operation of a business. General overhead is also associated with office, plant, equipment, staffing, and expenses thereof, maintained by a contractor for general business operations. The costs of general overhead are not specifically applicable to any given job or project (see Overhead).

Hammock—an aggregate or summary activity spanning the nodes of two or more activities and reported at a summary management level.

Hanger—a beginning or ending node not intended in the network (a break in a network path).

Histogram—see Block diagram.

Income—used interchangeably with profit. Avoid using Income instead of Sales Revenue. See Profit.

Incremental costs (benefits)—The additional cost (benefit) resulting from an increase in the investment in a project. Syn.: Marginal Cost (Benefit).

Indirect costs—(1) in construction, all costs that do not become a final part of the installation, but are required for the orderly completion of the installation and may include, but are not limited to, field administration, direct supervision, capital tools, startup costs, contractor's fees, insurance, taxes, etc; (2) In manufacturing, costs not directly assignable to the end product or process, such as overhead and general purpose labor, or costs of outside operations, such as transportation and distribution. Indirect manufacturing cost sometimes includes insurance, property taxes, maintenance, depreciation, packaging, warehousing and loading. In government contracts, indirect cost is often calculated as a fixed percent of direct payroll cost.

I-Node—(ADM) the node signifying the start of the activity (the tail of the arrow).

Input-output analysis—a matrix that provides a quantitative framework for the description of an economic unit. Basic to input-output analysis is a unique set of input-output ratios for each production and distribution process. If the ratios of input per unit of output are known for all production processes, and if the total production of each end product of the economy, or of the section being studied is known, it is possible to compute precisely the production levels required at every intermediate stage to supply the total sum of end products. Further, it is possible to determine the effect at every point in the production process of a specified change in the volume and mix of end products.

Interest—(1) financial share in a project or enterprise; (2) periodic compensation for the lending of money; (3) in economy study, synonymous with required return, expected profit, or charge for use of capital; (4) the cost for the use of capital. Sometimes referred to as the Time Value of Money.

Interest rate—the ratio of the interest payment to the principal for a given unit of time, usually expressed as a percentage of the principal.

Interest rate, compound—the rate earned by money expressed as a constant percentage of the unpaid balance at the end of the previous accounting period. Typical time periods are yearly, semiannually, monthly, and instantaneous.

Interest rate, continuous—see Compounding, continuous.

Interest rate, effective—an interest rate for a stated period (per year unless otherwise specified) that is the equivalent of a smaller rate of interest that is more frequently compounded.

Interest rate, nominal—the customary type of interest rate designation on an annual basis without consideration of compounding periods. A frequent basis for computing periodic interest payments.

Interface activity—an activity connecting an event in one subnetwork with an event in another subnetwork, and representing a logical or imposed inter-dependence between them.

Interface node—a common node for two or more subnets representing logical interdependence.

Intermediate node—a node where at least one activity begins and one activity ends.

Internal rate of return (IRR)—see Profitability index.

Inventory—raw materials, products in process, and finished products required for plant operation or the value of such material and other supplies, e.g., catalysts, chemicals, spare parts.

Investment—the sum of the original costs or values of the items that constitute the enterprise; used interchangeably with capital; may include expenses associated with capital outlays such as mine development.

Investment cost—includes first cost and later expenditures that have substantial and enduring value (generally more than one year) for upgrading, expanding, or changing the functional use of a facility, product, or process.

J-node—(ADM) the node signifying the finish of the activity (the head of the arrow).

Key activity—an activity that is considered of major significance. A key activity is sometimes referred to as a milestone activity.

Labor burden—taxes and insurances the employer is required to pay by law based on labor payroll, on behalf of or for the benefit of labor. (In the U.S.

these are federal old age benefits, federal unemployment insurance tax, state unemployment tax, and worker's compensation.)

Labor cost, manual—the salary plus all fringe benefits of construction workers and general labor on construction projects and labor crews in manufacturing or processing areas that can be definitely assigned to one product or process area or cost center.

Labor cost, nonmanual—in construction, normally refers to field personnel other than crafts and includes field administration and field engineering.

Labor factor—the ratio between the workhours actually required to perform a task under project conditions and the workhours required to perform an identical task under standard conditions.

Lead—a PDM constraint introduced before a series of activities to schedule them at a later time.

Learning curve—a graphic representation of the progress in production effectiveness as time passes. Learning curves are useful planning tools, particularly in the project oriented industries where new products are phased in rather frequently. The basis for the learning curve calculation is the fact that workers will be able to produce the product more quickly after they get used to making it.

Level float—the difference between the level finish and the imposed finish date.

Levelized fixed-charge rate—the ratio of uniform annual revenue requirements to the initial investment, expressed as a percent.

Life—(1) physical: that period of time after which a machine or facility can no longer be repaired in order to perform its design function properly. (2) service: the period of time that a machine or facility will satisfactorily perform its function without a major overhaul. See also Venture life; Economic life.

Life cycle—the stages, or phases that occur during the lifetime of an object or endeavor. A life cycle presumes a beginning and an end with each end implying a new beginning. In life cycle cost or investment analysis, the life cycle is the length of time over which an investment is analyzed (i.e., study period). see also Study period; Life.

Life cycle; asset—the stages, or phases of asset existence during the life of an asset. Asset life cycle stages typically include ideation, creation, operation, modification, and termination.

Life cycle; project—the stages or phases of project progress during the life of a project. Project life cycle stages typically include ideation, planning, execution, and closure.

Life-cycle cost (LCC) method—a technique of economic evaluation that sums over a given study period the costs of initial investment (less resale value), replacements, operations (including energy use), and maintenance and repair of an investment decision (expressed in present or annual value terms).

LIFO (Last In, First Out)—a method of determining the cost of inventory used in a product. In this method, the costs of material are transferred to the product in reverse chronological order. LIFO is used to describe the movement of goods.

Linear programming—mathematical techniques for solving a general class of optimization problems through minimization (or maximization) of a linear function subject to linear constraints. For example, in blending aviation fuel, many grades of commercial gasoline may be available. Prices and octane ratings, as well as upper limits on capacities of input materials which can be used to produce various grades of fuel are given. The problem is to blend the various commercial gasolines in such a way that (1) cost will be minimized (profit will be maximized), (2) a specified optimum octane rating will be met, and (3) the need for additional storage capacity will be avoided.

Load factor—(1) a ratio that applies to physical plant or equipment average load/maximum demand, usually expressed as a percentage. It is equivalent to percent of capacity operation if facilities just accommodate the maximum demand; (2) the ratio of average load to maximum load.

Load leveling—the technique of averaging, to a workable number, the amount or number of people working on a given project or in a given area of a project at a particular point in time. Load leveling is a benefit of most scheduling techniques and is necessary to insure a stable use of resources. Syn.: Workpower Leveling.

Location factor— an estimating factor used to convert the cost of an identical plant from one location to another. This factor takes into consideration the impact of climatic conditions, local infrastructure, local soil conditions, safety and environmental regulations, taxation and insurance regulations, labor availability and productivity, etc.

Logical restraint—a dummy arrow or constraint connection that is used as a logical connector but that does not represent actual work items. It is usually represented by a dotted line, and is sometimes called a dummy because it does not represent work. It is an indispensable part of the network concept when using the arrow diagramming method of CPM scheduling.

Loop—a path in a network closed on itself passing through any node or activity more than once, or, a sequence of activities in the network with no start or end.

Lump-sum—the complete in-place cost of a system, a subsystem, a particular item, or an entire project. Lump-sum contracts imply that no additional charges or costs will be assessed against the owner. see Contracts (2).

Maintenance and repair cost—the total of labor, material, and other related costs incurred in conducting corrective and preventative maintenance and repair on a facility, on its systems and components, or on both. Maintenance does not usually include those items that cannot be expended within the year purchased. Such items must be considered as fixed capital.

Management control systems—the systems (e.g., planning, scheduling, budgeting, estimating, work authorization, cost accumulation, performance measurement, etc.) used by owners, engineers, architects, and contractors to plan and control the cost and scheduling of work.

Management reserve—An amount added to an estimate to allow for discretionary management purposes outside of the defined scope of the project as otherwise estimated. Unlike contingency, the estimated reserve is not expected to be spent unless management so directs, and a reserve is generally not included in all estimates. An example of when a reserve might be included in a project estimate is when a project's schedule, safety, or operability are so critical to business objectives that business management authorizes reserve funds for project management to use at their discretion for any scope changes they feel are needed to meet the business objectives. Syns.: Reserve or Reserve Allowance.

Manufacturing cost—the total of variable and fixed or direct and indirect costs chargeable to the production of a given product, usually expressed in cents or dollars per unit of production, or dollars per year. Transportation and distribution costs, and research, development, selling and corporate administrative expenses are usually excluded. See also Operating cost.

Market value—the monetary price upon which a willing buyer and a willing seller in a free market will agree to exchange ownership, both parties knowing all the material facts but neither being compelled to act. The market value fluctuates with the degree of willingness of the buyer and seller and with the conditions of the sale. The use of the term market suggests the idea of barter. When numerous sales occur on the market, the result is to establish fairly definite market prices as the basis of exchanges.

Milestone—an important or critical event and/or activity that must occur when scheduled in the project cycle in order to achieve the project objective(s).

Milestone schedule—a schedule comprised of key events or milestones selected as a result of coordination between the client's and the contractor's project management. These events are generally critical accomplishments planned at time intervals throughout the project and used as a basis to monitor overall project performance. The format may be either network or bar chart and may contain minimal detail at a highly summarized level.

Monte Carlo method—a simulation technique by which approximate evaluations are obtained in the solution of mathematical expressions so as to determine the range or optimum value. The technique consists of simulating an experiment to determine some probabilistic property of a system or population of objects or events by use of random sampling applied to the components of the system, objects, or events.

Moving average—smoothing a time series by replacing a value with the mean of itself and adjacent values.

Multiple finish network—a network that has more than one finish activity or finish event.

Multiple start network—a network that has more than one start activity or event.

Multiple straight-line depreciation method—a method of depreciation accounting in which two or more straight line rates are used. This method permits a predetermined portion of the asset to be written off in a fixed number of years. One common practice is to employ a straight line rate which will write off 3/4 of the cost in the first half of the anticipated service life; with a second straight line rate to write off the remaining 1/4 in the remaining half life.

Near-critical activity—an activity that has low total float.

Net profit—earnings after all operating expenses (cash or accrued noncash) have been deducted from net operating revenues for a given period.

Net profit, percent of sales—the ratio of annual profits to total sales for a representative year of capacity operations. An incomplete measure of profitability, but a useful guidepost for comparing similar products and companies. Syn.: Profit margin.

Network—a logic diagram of a project consisting of the activities and events that must be accomplished to reach the objectives, showing their required sequence of accomplishments and interdependencies.

Network analysis—technique used in planning a project consisting of a sequence of activities and their interrelationship within a network of activities making up a project. See Critical path.

Node—the symbol on a logic diagram at the intersection of arrows (activities). Nodes identify completion and/or start of activities. See Event.

Nonexempt—employees not exempt from federal wage and hours guidelines.

Obsolescence—(1) the condition of being out of date. A loss of value occasioned by new developments that place the older property at a competitive disadvantage. A factor in depreciation; (2) a decrease in the value of an asset brought about by the development of new and more economical methods, processes, and/or machinery; (3) the loss of usefulness or worth of a product or facility as a result of the appearance of better and/or more economical products, methods or facilities.

Operating cost—the expenses incurred during the normal operation of a facility, or component, including labor, materials, utilities, and other related costs. Includes all fuel, lubricants, and normally scheduled part changes in order to keep a subsystem, system, particular item, or entire project functioning. Operating costs may also include general building maintenance, cleaning services, taxes, and similar items.

Operations research (OR)—quantitative analysis of industrial and administrative operations with intent to derive an integrated understanding of the factors controlling operational systems and in view of supplying management with an objective basis to make decisions. OR frequently involves representing the operation or the system with a mathematical model.

Overhead—a cost or expense inherent in the performing of an operation, i.e., engineering, construction, operating or manufacturing, which cannot be charged to or identified with a part of the work, product or asset and, therefore, must be allocated on some arbitrary base believed to be equitable, or handled as a business expense independent of the volume of production. Plant overhead is also called factory expense.

Over (under) plan—the planned cost to date minus the latest revised estimate of cost to date. When planned cost exceeds latest revised estimate, a projected underplan condition exists. When latest revised estimate exceeds planned cost, a projected overplan condition exists.

Overrun (underrun)—the value for the work performed to date minus the actual cost for that same work. When value exceeds actual cost, an underrun condition exists. When actual cost exceeds value, an overrun condition exists.

Parametric estimate—In estimating practice, describes estimating algorithms or cost estimating relationships that are highly probabilistic in nature (i.e., the parameters or quantification inputs to the algorithm tend to be abstractions of the scope). Typical parametric algorithms include, but are not limited to, factoring techniques, gross unit costs, and cost models (i.e., algorithms intended to replicate the cost performance of a process of system). Parametric estimates can be as accurate as definitive estimates.

Path—the logically continuous series of connected activities through a network.

Payout time—the time required to recover the original fixed investment from profit and depreciation. Most recent practice is to base payout time on an actual sales projection. Syn.: Payoff Period. See also Simple payback period.

Payroll burden—includes all payroll taxes, payroll insurances, fringe benefits, and living and transportation allowances.

PDM—see Precedence diagram method.

PDM arrow—a graphical symbol in PDM networks used to represent the lag describing the relationship between work items.

Percent complete—a comparison of the work completed to the current projection of total work. The percent complete of an activity in a program can be determined by inspection of quantities placed as workhours expended and compared with quantities planned or workhours planned. Other methods can also be used.

PERT—an acronym for Project Evaluation Review Technique, which is a probabilistic technique, used mostly by government agencies, for calculating the "most likely" durations for network activities. Most recently, however, the term PERT has been used as a synonym for CPM.

Phased construction—as most commonly used today, implies that construction of a facility or system or subsystem commences before final design is complete. Phased construction is used in order to achieve beneficial use at an advanced date.

Plant overhead—those costs in a plant that are not directly attributable to any one production or processing unit and are allocated on some arbitrary basis believed to be equitable. Includes plant management salaries, payroll department, local purchasing and accounting, etc. Syn.: Factory Expense.

Precedence diagram method (PDM)—a method of constructing a logic network using nodes to represent the activities and connecting them by lines that show logic relationships.

Preconstruction CPM—a plan and schedule of the construction work developed during the design phase preceding the award of contract.

Price index—the representation of price changes, which is usually derived by dividing the current price for a specific good by some base period price. See Cost index.

Pricing—In estimating practice, after costing an item or activity, the determination of the amount of money asked in exchange for the item, activity, or project. Pricing determination considers business and other interests (e.g., profit, marketing, etc.) in addition to inherent costs. The price may be greater or less than the cost depending on the business or other objectives. In the cost estimating process, pricing follows costing and precedes budgeting. In accounting practice, the observation and recording (collecting) of prices.

Probability—a value depicting the likelihood of an expected or occurred event.

Profit—(1) Gross Profit—earnings from an ongoing business after direct costs of goods sold have been deducted from sales revenue for a given period. (2) Net Profit—earnings or income after subtracting miscellaneous income and expenses (patent royalties, interest, capital gains) and federal income tax from operating profit. (3) Operating Profit—earnings or income after all expenses (selling, administrative, depreciation) have been deducted from gross profit.

Profitability—a measure of the excess income over expenditure during a given period of time.

Profitability analysis—the evaluation of the economics of a project, manufactured product, or service within a specific time frame.

Profitability index (PI)—the rate of compound interest at which the company's outstanding investment is repaid by proceeds for the project. All proceeds from the project, beyond that required for interest, are credited, by the method of solution, toward repayment of investment by this calculation. Also called discounted cash flow, interest rate of return, investor's method, internal rate of return. Although frequently requiring more time to calculate than other valid yardsticks, PI reflects in a single number both the dollar and the time values of all money involved in a project. In some very special cases, such as multiple changes of sign in cumulative cash position, false and multiple solutions can be obtained by this technique.

Projection—an extension of a series, or any set of values, beyond the range of the observed data.

Project management—the utilization of skills and knowledge in coordinating the organizing, planning, scheduling, directing, controlling, monitoring and evaluating of prescribed activities to ensure that the stated objectives of a project, manufactured product, or service, are achieved.

Project manager—an individual who has been assigned responsibility and authority for accomplishing a specifically designated unit of work effort or group of closely related efforts established to achieve stated or anticipated objectives, defined tasks, or other units of related effort on a schedule for performing the stated work funded as a part of the project. The project manager is responsible for the planning, controlling, and reporting of the project.

Project network analysis (PNA)—a group of techniques based on the network project representation to assist managers in planning, scheduling, and controlling a project.

Project summary work breakdown structure (PSWBS)—a summary WBS tailored by project management to the specific project, and identifying the elements unique to the project.

Punchlist—a list generated by the owner, architect, engineer, or contractor of items yet to be completed by the contractor. Sometimes called a "but" list ("but" for these items the work is complete).

Quality—conformance to established requirements (not a degree of goodness).

Quality acceptance criteria—specified limits placed on characteristics of a product, process, or service defined by codes, standards, or other requirement documents.

Quality control—inspection, test, evaluation or other necessary action to verify that a product, process, or service conforms to established requirements and specifications.

Quantity survey—using standard methods measuring all labor and material required for a specific building or structure and itemizing these detailed quantities in a book or bill of quantities.

Quantity surveyor—In the United Kingdom, contractors bidding a job receive a document called a bill of quantities, in addition to plans and specifications, which is prepared by a quantity surveyor, according to well-established rules. To learn these rules the quantity surveyor has to undergo five years of technical training and must pass a series of professional examinations. In the United Kingdom a quantity surveyor establishes the quantities for all bidders, and is professionally licensed to do so.

Queuing theory—the theory involving the use of mathematical models, theorems and algorithms in the analysis of systems in which some service is to be performed under conditions of randomly varying demand, and where waiting lines or queues may form due to lack of control over either the demand for service or the amount of service required, or both. Utilization of the theory extends to process, operation and work studies.

Random process—in a general sense the term is synonymous with the more usual and preferable "stochastic" process. It is sometimes employed to denote a process in which the movement from one state to the next is determined by a variate independent of the initial and final state.

Random walk—the path traversed by a particle that moves in steps, each step being determined by chance either in regard to direction or in regard to magnitude or both. Cases most frequently considered are those in which the particle moves on a lattice of points in one or more dimensions, and at each step is equally likely to move to any of the nearest neighboring points. The theory of random walks has many applications, e.g., to the migration of insects, sequential sampling and, in the limit, to diffusion processes.

Rate of return—the interest rate earned by an investment. See Return on average investment, Return on original investment, Profitability index, Discounted cash flow.

Real property—refers to the interests, benefits, and rights inherent in the ownership of physical real estate. It is the bundle of rights with which the ownership of real estate is endowed.

Regression—a functional relationship between two or more correlated variables often empirically determined from data and used to predict values of one variable when given values of the others.

Replacement cost—(1) the cost of replacing the productive capacity of existing property by another property of any type, to achieve the most economical service, at prices as of the date specified; (2) facility component replacement and related costs, included in the capital budget, that are expected to be incurred during the study period.

Replacement value—that value of an item determined by repricing the item on the basis of replacing it, in new condition, with another item that gives the same ability to serve, or the same productive capacity, but which applies current economic design, adjusted for the existing property's physical deterioration.

Resource—in planning and scheduling, a resource is any consumable, except time, required to accomplish an activity. From a total cost and asset management perspective, resources may include any real or potential investment in strategic assets including time, monetary, human, and physical. A resource becomes a cost when it is invested or consumed in an activity or project.

Resource allocation process (RAP)—the scheduling of activities in a network with the knowledge of certain resource constraints and requirements. This process adjusts activity level start and finish dates to conform to resource availability and use.

Resource code—the code for a particular labor skill, material, equipment type; the code used to identify a given resource.

Resource limited scheduling—a schedule of activities so that a preimposed resource availability level (constant or variable) is not exceeded in any given project time unit.

Return on average investment—the ratio of annual profits to the average book value of fixed capital, with or without working capital. This method has some advantages over the return-on-original-investment method. Depreciation is always considered; terminal recoveries are accounted for. However, the method does not account for the timing of cash flow and yields answers that are considerably higher than those obtained by the return-on-original-investment and profitability index methods. Results may be deceiving when compared, say, against the company's cost of capital.

Return on original investment—the ratio of expected average annual after tax profit (during the earning life) to total investment (working capital included). It is similar in usefulness and limitations to payoff period.

Risk—the degree of dispersion or variability around the expected or "best" value which is estimated to exist for the economic variable in question, e.g., a quantitative measure of the upper and lower limits considered reasonable for the factor being estimated.

Salvage value— (1) the cost recovered or which could be recovered from a used property when removed, sold, or scrapped; (2) the market value of a machine or facility at any point in time (normally an estimate of an asset's net market value at the end of its estimated life); (3) the value of an asset, assigned for tax computation purposes, that is expected to remain at the end of the depreciation period.

Schedule variance—the difference between BCWP and BCWS. At any point in time it represents the difference between the dollar value of work actually performed (accomplished) and that scheduled to be accomplished.

Scope—The sum of all that is to be or has been invested in and delivered by the performance of an activity or project. In project planning, the scope is usually documented (i.e., the scope document), but it may be verbally or otherwise communicated and relied upon. Generally limited to that which is agreed to by the stakeholders in an activity or project (i.e., if not agreed to, it is out of scope.). In contracting and procurement practice, includes all that an enterprise is contractually committed to perform or deliver. Syn.: Project Scope.

Scope change—a deviation from the project scope originally agreed to in the contract. A scope change can consist of an activity either added to or deleted from the original scope. A contract change order is needed to alter the project scope.

Secondary float (SF)—is the same as the Total Float, except that it is calculated from a schedule date set upon an intermediate event.

Shutdown point—the production level at which it becomes less expensive to close the plant and pay remaining fixed expenses out-of-pocket rather than continue operations; that is, the plant cannot meet its variable expense. See Breakeven point.

Simple interest—(1) interest that is not compounded—is not added to the income-producing investment or loan; (2) the interest charges under the condition that interest in any time period is only charged on the principal.

Simulation—(1) the technique of utilizing representative or artificial operating and demand data to reproduce, under test, various conditions that are likely to occur in the actual performance of a system. Simulation is frequently used to test the accuracy of a theoretical model or to examine the behavior of a system under different operating policies; (2) The design and operation of a model of a system.

Sinking fund—(1) a fund accumulated by periodic deposits and reserved exclusively for a specific purpose, such as retirement of a debt or replacement of a property; (2) a fund created by making periodic deposits (usually equal) at compound interest in order to accumulate a given sum at a given future time for some specific purpose.

Slack paths—the sequences of activities and events that do not lie on the critical path or paths.

Slack time—the difference in calendar time between the scheduled due date for a job and the estimated completion date. If a job is to be completed ahead of schedule, it is said to have slack time; if it is likely to be completed behind schedule, it is said to have negative slack time. Slack time can be used to calculate job priorities using methods such as the critical ratio. In the critical path method, total slack is the amount of time a job may be delayed in starting without necessarily delaying the project completion time. Free slack is the amount of time a job may be delayed in starting without delaying the start of any other job in the project.

Standard deviation—the most widely used measure of dispersion of a frequency distribution. It is calculated by summing squared deviations from the mean, dividing by the number of items in the group and taking the square root of the quotient.

Standard network diagram—a predefined network intended to be used more than one time in any given project.

Startup—that period after the date of initial operation, during which the unit is brought up to acceptable production capacity and quality within estimated production costs. Startup is the activity that commences on the date of initial activity that has significant duration on most projects, but is often confused (used interchangeably) with date of initial operation.

Startup costs—extra operating costs to bring the plant on stream incurred between the completion of construction and beginning of normal operations. In addition to the difference between actual operating costs during that period and normal costs, it also includes employee training, equipment tests, process adjustments, salaries and travel expense of temporary labor, staff and consultants, report writing, poststartup monitoring and associated overhead. Additional capital required to correct plant problems may be included. Startup costs are sometimes capitalized.

Stochastic—the adjective "stochastic" implies the presence of a random variable.

Straight-line depreciation—method of depreciation whereby the amount to be recovered (written off) is spread uniformly over the estimated life of the asset in terms of time periods or units of output.

Study period—the length of time over which an investment is analyzed. Syn: Life cycle; time horizon.

Sum-of-digits method—A method of computing depreciation in which the amount for any year is based on the ratio: (years of remaining life)/$(1 + 2 + 3 + \cdots + n)$, n being the total anticipated life. Also known as sum-of-the-years-digits method.

Sunk cost—a cost that has already been incurred and should not be considered in making a new investment decision.

Takeoff—measuring and listing from drawings the quantities of materials required in order to price their cost of supply and installation in an estimate and to proceed with procurement of the materials.

Time-scaled CPM—a plotted or drawn representation of a CPM network where the length of the activities indicates the duration of the activity as drawn to a calendar scale. Float is usually shown with a dashed line as are dummy activities.

Total float (TF)—the amount of time (in work units) that an activity may be delayed from its early start without delaying the project finish date. Total float is equal to the late finish minus the early finish or the late start minus the early start of the activity.

Unit cost—dollar per unit of production. It is usually total cost divided by units of production, but a major cost divided by units of production is frequently referred to as a unit cost; for example, the total unit cost is frequently subdivided into the unit costs for labor, chemicals, etc.

Value engineering—a practice function targeted at the design itself, which has as its objective the development of design of a facility or item that will yield least life-cycle costs or provide greatest value while satisfying all performance and other criteria established for it.

Variable costs—those costs that are a function of production, e.g., raw materials costs, byproduct credits, and those processing costs that vary with plant output (such as utilities, catalysts and chemical, packaging, and labor for batch operations).

Variance—in cost control, the difference between actual cost or forecast budget cost.

Venture life—the total time span during which expenditures and/or reimbursements related to the venture occur. Venture life may include the research and development, construction, production and liquidation periods.

Venture worth—present worth of cash flows above an acceptable minimum rate, discounted at the average rate of earnings.

Weights—numerical modifiers used to infer importance of commodities in an aggregative index.

Work—any and all obligations, duties, responsibilities, labor, materials, equipment, temporary facilities, and incidentals, and the furnishing thereof necessary to complete the construction which are assigned to, or undertaken by the contractor, pursuant to the contract documents. Also, the entire completed construction or the various separately identifiable parts thereof required to be furnished under the contract documents. Work is the result of performing services, furnishing labor, and furnishing and incorporating materials and equipment into the construction, all as required by the contract documents.

Work breakdown structure (WBS)—a product-oriented family tree division of hardware, software, facilities and other items which organizes, defines and displays all of the work to be performed in accomplishing the project objectives.

1. Contract Work Breakdown Structure (CWBS)—the complete WBS for a contract developed and used by a contractor in accordance with the contract work statement. It extends the PSWBS to the lowest level appropriate to the definition of the contract work.
2. Project Summary Work Breakdown Structure (PSWBS)—a summary WBS tailored by project management to the specific project with the addition of the elements unique to the project.

Work-in-process—product in various stages of completion throughout the factory, including raw material that has been released for initial processing and completely processed material awaiting final inspection and acceptance as finished product or shipment to a customer. Many accounting systems also include semifinished stock and components in this category. Syn: In-process inventory.

Work package—a segment of effort required to complete a specific job such as a research or technological study or report, experiment or test, design specification, piece of hardware, element of software, process, construction drawing, site survey, construction phase element, procurement phase element, or service, within the responsibility of a single unit within the performing organization. The work package is usually a functional division of an element of the lowest level of the WBS. Syn: In-process Inventory.

Worth—the worth of an item or groups of items, as in a complete facility, is determined by the return on investment compared to the amount invested. The worth of an item is dependent upon the analysis of feasibility of the entire item or group or items under discussion (or examination).

Yield—the ratio of return or profit over the associated investment, expressed as a percentage or decimal usually on an annual basis. See Rate of return.

Appendix C

Guide to Project and Cost Related Internet Sites

C.1 INTRODUCTION

The Internet is an invaluable source of information for cost engineers and project managers, if you know where to look. This appendix provides many different Internet addresses of value. The listing is far from complete but does provide a representative sample of the major web sites in each listed category. The addresses provided are those felt to have a reasonable probability of not changing in the foreseeable future. However, the Internet is very fluid and it is probable that at least a few of these addresses will change, if for no other reason than corporate mergers or changes of corporate names in coming years. Should any address change and no longer be valid, conduct a search for updated address information using Google (http://www.google.com) or any of the other major Internet search engines.

C.2 COST ENGINEERING, QUANTITY SURVEYING, AND PROJECT MANAGEMENT ASSOCIATIONS AND INSTITUTES

International Confederations of Associations and Institutes

> International Cost Engineering Council—http://www.icoste.org
> International Federation of Surveyors—http://www.fig.net
> International Project Management Association—http://www.ipma.ch
> Pacific Association of Quantity Surveyors—http://www.paqs.net

National Associations
Australia

> Australian Institute of Quantity Surveyors—http://www.aiqs.com.au

Austria

Projekt Management Austria (Austrian Project Management Association)—http://www.p-m-a.at

Azerbaijan

Azerbaijan Project Management Association—http://www.azpma.net

Brazil

Instituto Brasileiro de Engenharia de Costos—http://www.ibec.org.br

Canada

AACE Canada, The Canadian Association of Cost Engineers—http://www.aacei.org
Association of Quantity Surveyors of Alberta—http://www.qs-alta.org
Canadian Institute of Quantity Surveyors—http://www.ciqs.org
Ontario Institute of Quantity Surveyors—http://www.oiqs.org
Quantity Surveyors Society of British Columbia—http://www.qssbc.org

China

China Engineering Cost Association—http://www.ceca.org.cn

Czech Republic

Spolocnost pro projektové rízení (Czech Association of Project Management)—http://www.ipma.cz

Denmark

Foreningen for Dansk Projektledelse (Danish Project Management Association)—http://www.projektforeningen.dk

Egypt

Management Engineering Society—http://www.mes.eg.net

Finland

Projektiyhdistys (Project Management Association of Finland)—http://www.pry.fi

France

Association Francophone de Management de Projet (Project Management Association of France)—http://www.afitep.fr

Germany

Deutsche Gesellschaft für Projektmanagement (Project Management Association of Germany)—http://www.gpm-ipma.de

Greece

Hellenic Institute for Project Management—http://www.hpmi.gr

Hong Kong

Hong Kong Institute of Surveyors—http://www.hkis.org.hk

Hungary

Project Management Association of Hungary—http://www.fovosv.hu

Iceland

Verkefnastjornunarfelag (Project Management Association of Iceland)—
http://www.vsf.is

India

Project Management Associates—India—http://www.pma-india.org

Ireland

Institute of Project Management of Ireland—http://www.projectmanage-
ment.ie

Italy

Associazione Italiana d'Ingegneria Economica (Italian Association of Cost
Engineers)—http://www.aice-it.org

Japan

Japan Project Management Forum—http://www.enaa.or.jp/JPMF
Nihon Kenchiku Sekisan Kyokai (Building Surveyors Institute of Japan)—
http://www.bsij.or.jp

Kenya

Institute of Quantity Surveyors of Kenya—http://www.iqsk.org

Korea

Korean Institute of Project Management & Technology—http://www.
promat.or.kr

Latvia

Latvian National Project Management Association—http://www.
lnpva.lv

Malaysia

Pertubuhan Juruukur Malaysia (Institution of Surveyors, Malaysia)—
http://www.ism.org.my

Mexico

Sociedad Mexicana de Ingenieria Economica, Financiera y de Costos (Mexican Society of Engineering Economics, Financing, and Costs)— http://www.smiefc.com.mx

Netherlands

Nederlandse Stichting Voor Kostentechniek (Dutch Association of Cost Engineers)—http://www.napdace.nl
PMI-Nederland (Project Management Association of the Netherlands)— http://www.pmi-nl.nl

New Zealand

New Zealand Institute of Quantity Surveyors—http://www.nziqs.co.nz

Nigeria

Nigerian Institute of Quantity Surveyors—http://www.niqs.org

Norway

Norsk Forening for Prosjektledelse (Norwegian Project Management Association)—http://www.prosjektledelse.com

Poland

Stowarzyszenie Project Management Polska—http://www.spmp.org.pl

Portugal

Associaçao Portuguesa Gestao de Projectos (Portuguese Project Management Association)—http://www.apogep.pt

Romania

Project Management Romania—http://www.pm.org.ro

Russia

Russian Association of Bidders and Cost Engineering—http://www.rabce.da.ru
SOVNET (Russian Project Management Association)—http://www.sovnet.ru

Slovakia

Spolocnost pre projektove riadenie (Project Management Association of Slovakia)—http://www.sppr.sk

Slovenia

Slovenian Project Management Association—http://www.cd-cc.si/ZPM

Singapore

Singapore Institute of Surveyors and Valuers—http://www.sisv.org.sg

South Africa

Association of South African Quantity Surveyors—http://www.asaqs. co.za
Cost Engineering Association of Southern Africa—http://www.ceasa. org.za
Project Management South Africa—http://www.pmisa.co.za

Spain

Asociación Española de Ingeniería de Proyectos (Spanish Project Engineering Association)—http://www.aeipro.org

Sweden

Svenskt ProjektForum (Swedish Project Management Association)— http://www.projforum.se

Switzerland

Schweizerische Gesellschaft für Projektmanagement (Project Management Association of Switzerland)—http://www.spm.ch

Ukraine

Ukranian Project Management Association—http://www.freenet.kiev.ua

United Kingdom

Association for Project Management—http://www.apm.org.uk
Association of Cost Engineers—http://www.acoste.org.uk
Royal Institution of Chartered Surveyors—http://www.rics.org.uk

United States of America

AACE International, The Association for the Advancement of Cost Engineering—http://www.aacei.org
American Society of Professional Estimators—http://www.aspenational.com
Construction Specifications Institute—http://www.csinet.org
International Society of Parametric Analysts—http://www.ispa-cost.org
Project Management Institute—http://www.pmi.org
SAVE International, The Value Society—http://www.value-eng.org
Society of Cost Estimating and Analysis—http://www.sceaonline.net

Venezuela

Guia Referencial de Costos (Venezuela Cost Reference Guide)—http:// www.grc.com.ve

Yugoslavia

YUPMA (Yugoslov Project Management Association)—http://www.fon. bg.ac.yu/yupma

C.3   COST ENGINEERING AND PROJECT MANAGEMENT SOFTWARE

The following listing is divided into the broad categories of (1) cost estimating software, (2) project management, planning and scheduling software, and (3) risk analysis and range estimating software. So much software is available that it would be impossible to list every supplier and software package. In an effort to be as representative as possible of software suppliers to the cost engineering profession, a study was made of the firms that regularly advertised in *Cost Engineering* magazine during 2002 and 2003. The following listings include those advertisers plus a limited number of others with which the editor of this book is familiar. Inclusion of any firm or software package in this listing does not necessarily constitute a recommendation or endorsement. Omission of any firm or software package similarly should not be construed in a negative manner.

C.3.1   Cost Estimating Software

ARES Corporation. Provider of *PRISM* software products including *PRISM Project Estimator*—http://www.arescorporation.com

Building Systems Design, Inc. Provider of *CostLink*®/*CM* cost estimating software—http://www.bsdsoftlink.com

Cost Xpert Group, Inc. Provider of software cost estimating tools, training, and consulting—http://www.costxpert.com

CPR International, Inc. Provider of *Visual Estimator*™ construction cost estimating and bidding system; *GeneralCOST Estimator*™ *for Excel* general construction cost estimating system for *MS Excel*; and *Plan Takeoff*™ construction plan takeoff system—http://www.cprsoft.com

Day Bennett & Associates. Distributor of *Costrac*™ software for project cost management—http://www.costrac.net

Estimating4U. Provider of *Estimating4u.com*, a cost estimating tool for quickly creating project estimates—http://www.estimating4u.com

InSite Software. Provider of earthwork and utility estimating software—http://www.insitesoftware.com

International Project Estimating Software. An estimating system for major international infrastructure and internationally tendered projects—http://www.ipestimate.com

Pronamics Pty. Ltd. Suppliers of elemental estimating software for engineering and construction—http://www.pronamics.com.au

Quest Solutions, Inc. Provider of earth-measuring and estimating software for the construction industry—http://www.questsolutions. com

Timberline Software Corporation. A leading provider of estimating and other software for construction and other industries—http://www. timberline.com

U.S. COST. Provider of *SuccessEstimator* software for cost estimating and management—http://www.uscost.com

WinEstimator Inc.® Provider of *WinEst*® software for project cost estimating—http://www.winest.com

C.3.2   Project Management, Planning and Scheduling Software

American eBusiness Solutions. Provider of *EPMAC* web-based enterprise solution for project management and team collaboration—http:// www.epmac.com

ARES Corporation. Provider of *PRISM* software products including *PRISM Project Manager*—http://www.arescorporation.com

Axista.com. Provider of web based project management software for team collaboration—http://www.axista.com

Computer Guidance Corporation. Provider of project management, construction management, and bid solicitation software—http://www. computer-guidance.com

Integrated Management Concepts Inc. Provider of *MicroFusion Millennium* cost and earned value management software—http://www. intgconcepts.com

InterPlan Systems Inc. Provider of *eTaskMaker*™ project planning software and *ATC Professional*™ project management software (for oil refining and petrochemical plant maintenance turnarounds)—http:// www.interplansystems.com

Kildrummy, Inc. Provider of *CostMANAGER* project controls software— http://www.kildrummy.com

Meridian Project Systems. Provider of project management software systems—http://www.mps.com

OnTrack Engineering Ltd. Provider of *CostTrack*™ integrated project controls software package

Primavera Systems, Inc. A leading provider of cost and schedule project management software—http://www.primavera.com

Rational Concepts Inc. Provider of Internet-based cost/schedule project management software—http://www.rationalconcepts.com

Ron Winter Consulting. Provider of *Schedule Analyzer* schedule review software—http://www.ronwinterconsulting.com

TimeTraXX Software. Provider of *TimeTraXX*™ software for time studies and schedule impact analysis—http://timetraxxsoftware.com

Welcom Software Technology. A leading provider of cost and schedule project management software—http://www.welcom.com

C.3.3 Risk Analysis and Range Estimating Software

Decision Sciences Corporation. Developer of *REP/PC* software for range estimating and Monte Carlo simulation—http://uncertain.com

Decisioneering®. Distributor of *Crystal Ball®* risk analysis software—http://www.decisioneering.com

Palisade Corporation. Distributor of software and publications in risk analysis, decision analysis, and optimization including @*RISK* software for risk analysis

C.4 PUBLISHERS OF COST DATA AND INFORMATION

As was the case for the listings above of software suppliers, it would be impossible to provide a fully comprehensive listing of publishers of books, databases and information sources that are available to the cost engineering profession. The list below is far from complete but it does include many of the more prominent publishers in the field. It includes publishers of electronic information as well as publishers of traditional printed material. In addition, all of the associations and institutes listed in Section C.1 of this appendix publish books, journals, newsletters, and conference proceedings.

Architectural Record. Online resource of the architectural community with continuing eduction courses, news, buyer's guide, etc.—http://archrecord.construction.com

Architects' First Source. A building product information source. Producer of CSI's *SPEC-DATA®* and *MANU-SPEC®* and *CADBlocks* to deliver manufacturer's technical product data, proprietary specifications, and drawings—http://www.firstsourceonl.com

Aspen Technology/Richardson Engineering Services. Electronic publisher of: *Aspen Richardson's General Construction Cost Estimating Standards, Aspen Richardson's Process Plant Construction Cost Estimating Standards*, and *Aspen Richardson's International Construction Factors Location Cost Manual*—http://www.aspentech.com/resi.cfm

Associated Construction Publications. Publisher of 14 regional magazines focused on highway and heavy construction and cost data—http://www.acppubs.com

The Blue Book of Building and Construction. Directory of regional, categorized construction information and services—http://www.thebluebook.com

Byggfakta Scandinavia AB. Provider of construction project information for Denmark, Finland, Norway and Sweden—http://www.bf-scandinavia.com

Clark Reports. Provider of industrial project information for North America—http://www.clarkreports.com

ConstructionRisk.Com. An online construction risk management library—http://www.constructionrisk.com

ConstructionWebLinks.com. A guide to construction resources on the Internet. ConstructionWebLinks.com indexes, profiles, and links to more than 2,500 different web sites with information useful to construction industry professional—http://www.constructionweblinks.com

Design Build Magazine. Information on project delivery via the design-build method of construction—http://designbuild.construction.com

Earned Value Management. Information on earned value project management (US Department of Defense web site)—http://www.acq.osd.mil/pm

EC Harris Capital Projects and Facilities Consultants. Online publisher of *Global Review* which provides comparative building construction costs, tax rate information, GDP forecasts, and retail price inflation forecasts for many countries—http://www.echarris.com/publications

Edinburg Engineering Virtual Library. Searchable database of engineering information—http://www.eevl.ac.uk

Engineering News Record Magazine. Weekly news magazine serving the building and construction industry. Compiler of building and construction cost indexes. Frequent articles concerning international costs and a special quarterly cost edition—http://enr.construction.com

Frank R. Walker Company Publisher of the *Building Estimator's Reference Book* - http://www.frankrwalker.com

FW Dodge.Com: The Construction Project Marketplace. A searchable online database of over 500,000 construction projects worldwide—http://www.fwdodge.com

The Hampton Group. Site offers free project management articles, book chapters, a free monthly newsletter, and project management training courses—http://www.projectmanagertraining.com/articles_pm.html

Hanscomb Faithful & Gould. Co-publisher of *The Hanscomb/Means Report: International Construction Cost Intelligence*—http://www.hanscombfgould.com

Marcel Dekker, Inc. Publisher of numerous books on cost engineering and project management topics—http://www.dekker.com

Marshall and Swift. Publisher of construction cost data and compiler of cost indexes—http://www.marshallswift.com

McGraw-Hill Construction Information Group. Construction news and links to various sources of construction and cost information—http://www.construction.com

Norsk Byggtjeneste A. S. (Norwegian Building Center). Maintains a database of prices for thousands of products—http://www.byggtjeneste.no/english.htm

Norwegian Statistical Bureau (Statistic Sentralbyrå). Maintains a database economic and statistical information for Norway. Publishes *Monthly Bulletin of Statistics* which contains construction cost information— http://www.ssb.no/english

Planning Planet. A database and discussion forum for construction project planners and managers—http://www.planningplanet.com

The Rawlinsons Group. Publisher of cost reference books for Australia and the Asia-Pacific region—http://www.rawlinsons.com.au

Reed Construction Data. A group of companies providing data and information services for the construction industry—http://www.reedconstructiondata.com

R. S. Means Co. A leading publisher of construction cost data books and reference books—http://www.rsmeans.com

Spon Press. Publisher of numerous cost reference books—http://www.efnspon.com

Steelonthenet.Com. Steel industry web site providing global steel pricing information and industry news—http://www.steelonthenet.com

Sweets.com: The Construction Product Marketplace. Industry and product information, specifications and drawings on over 25,000 building products and built-in RFQ/RFP functions—http://sweets.construction.com

C.5   OTHER WEB SITES OF INTEREST

The following is a listing of miscellaneous other Internet sites that may be of interest to cost and project professionals.

Federal Reserve Bank of New York. Site provides foreign exchange rates and other financial information—http://www.ny.frb.org/markets

National Aeronautics and Space Administration. Page provides numerous sources of cost and wage information. This page is particularly useful for those involved in work for the United States Federal Government— http://www.jsc.nasa.gov/bu2/data.html

Norwegian Center for Project Management. Sponsor of research on management by projects, virtual project organizations, managing external relations, and project management in practice—http://www.nsp.ntnu.no/default.asp?lng=english

OANDA.com: The Currency Site. A multilingual currency converter which converts currency amounts for 164 countries at the exchange rates on any given day—http://www.oanda.com/converter/classic

ProVoc Project Control Engineers: National Vocational Qualifications. ProVoc is a steering Committee raising the profile of United Kingdom National and Scottish Vocational Qualifications (N/SVQ) for Project Management and Project Control staff in industry and commerce— http://www.provoc.org.uk

U. S. Bureau of Labor Statistics Producer Price Indexes. Site provides "Producer Price Indexes" for commodities and industrial products—http://www.bls.gov/pp

U. S. Bureau of Labor Statistics Foreign Labor Statistics Site. provides labor cost information and statistics for many countries—http://www.bls. gov/fls

Index